PLANE TRIGONOMETRY
PART-II

PLANE TRIGONOMETRY
PART-II

S L LONEY
Former Professor of Mathematics
Royal Holloway College, Egham Surrey, UK
Fellow, Sidney Sussex College, Cambridge, UK

New Academic Science
An Imprint of
New Age International (UK) Ltd.
27 Old Gloucester Street, London, WC1N 3AX, UK
www.newacademicscience.co.uk • e-mail: info@newacademicscience.co.uk

Copyright © 2018 by New Academic Science Limited

27 Old Gloucester Street, London, WC1N 3AX, UK

www.newacademicscience.co.uk • email: info@newacademicscience.co.uk

---

ISBN: 978 1 78183 110 6

British Library Cataloguing in Publication Data

A Catalogue record for this book is available from the British Library

Every effort has been made to make the book error free. However, the author and publisher have no warranty of any kind, expressed or implied, with regard to the documentation contained in this book.

PUBLISHER'S NOTE

This book has been thoroughly read by our subject experts and efforts have been made to present an error free edition. Though the overall format is not disturbed but the presentation and the rubric has been amplified. More clarity is incorporated wherever necessary and the figures are redrawn for better understanding. Suggestions for further improvement shall be received with appreciation.

Contents

The Principal Formulae in Trigonometry

I. Circumference of a circle= $2\pi r$. (Art. 12)

$\pi = 3.14159...$ $\left[\text{Approximations are } \dfrac{22}{7} \text{ and } \dfrac{355}{113}\right]$. (Art. 13)

A Radian = $57° \ 17' \ 44.8''$ nearly. (Art. 16)

Two right angles = $180° = 200^g = \pi$ radians. (Art. 19)

$\text{Angle} = \dfrac{\text{arc}}{\text{radius}} \times \text{Radian}$. (Art. 21)

II. $\sin^2\theta + \cos^2\theta = 1$;

$\sec^2\theta = 1 + \tan^2\theta$;

$\operatorname{cosec}^2\theta = 1 + \cot^2\theta$. (Art. 27)

III. $\sin 0° = 0$; $\cos 0° = 1$. (Art. 36)

$\sin 30° = \dfrac{1}{2}$; $\cos 30° = \dfrac{\sqrt{3}}{2}$. (Art. 34)

$\sin 45° = \cos 45° = \dfrac{1}{\sqrt{2}}$. (Art. 33)

$\sin 60° = \dfrac{\sqrt{3}}{2}$; $\cos 60° = \dfrac{1}{2}$. (Art. 35)

$\sin 90° = 1$; $\cos 90° = 0$. (Art. 37)

$\sin 15° = \dfrac{\sqrt{3}-1}{2\sqrt{2}}$; $\cos 15° = \dfrac{\sqrt{3}+1}{2\sqrt{2}}$. (Art. 106)

$\sin 18° = \dfrac{\sqrt{5}-1}{4}$; $\cos 36° = \dfrac{\sqrt{5}+1}{4}$. (Arts. 120, 121)

IV. $\sin(-\theta) = -\sin\theta$; $\cos(-\theta) = \cos\theta$. (Art. 68)

$\sin(90° - \theta) = \cos\theta$; $\cos(90° - \theta) = \sin\theta$. (Art. 69)

$\sin(90° + \theta) = \cos\theta$; $\cos(90° + \theta) = -\sin\theta$. (Art. 70)

$\sin(180° - \theta) = \sin\theta$; $\cos(180° - \theta) = -\cos\theta$. (Art. 72)

$\sin(180° + \theta) = -\sin\theta; \cos(180° + \theta) = -\cos\theta.$ (Art. 73)

V.　If $\sin\theta = \sin\alpha$, then $\theta = n\pi + (-1)^n\alpha$ (Art. 82)

　　If $\cos\theta = \cos\alpha$, then $\theta = 2n\pi \pm \alpha.$ (Art. 83)

　　If $\tan\theta = \tan\alpha$, then $\theta = n\pi + \alpha.$ (Art. 84)

VI.　$\sin(A + B) = \sin A \cos B + \cos A \sin B.$

　　$\cos(A + B) = \cos A \cos B - \sin A \sin B.$ (Art. 88)

　　$\sin(A - B) = \sin A \cos B - \cos A \sin B.$

　　$\cos(A - B) = \cos A \cos B + \sin A \sin B.$ (Art. 90)

$$\sin C + \sin D = 2\sin\frac{C+D}{2}\cos\frac{C-D}{2}.$$

$$\sin C - \sin D = 2\cos\frac{C+D}{2}\sin\frac{C-D}{2}.$$

$$\cos C + \cos D = 2\cos\frac{C+D}{2}\cos\frac{C-D}{2}.$$

$$\cos D - \cos C = 2\sin\frac{C+D}{2}\sin\frac{C-D}{2}.$$ (Art. 94)

$2\sin A \cos B = \sin(A + B) + \sin(A - B).$

$2\cos A \sin B = \sin(A + B) - \sin(A - B).$

$2\cos A \cos B = \cos(A + B) + \cos(A - B).$

$2\sin A \sin B = \cos(A - B) - \cos(A + B).$ (Art. 97)

$$\tan(A + B) = \frac{\tan A + \tan B}{1 - \tan A \tan B}.$$

$$\tan(A - B) = \frac{\tan A - \tan B}{1 + \tan A \tan B}.$$ (Art. 98)

$\sin 2A = 2\sin A \cos A.$

$\cos 2A = \cos^2 A - \sin^2 A = 1 - 2\sin^2 A = 2\cos^2 A - 1.$

$$\sin 2A = \frac{2\tan A}{1 + \tan^2 A}; \quad \cos 2A = \frac{1 - \tan^2 A}{1 + \tan^2 A}.$$ (Art. 109)

$$\tan 2A = \frac{2\tan A}{1 - \tan^2 A}.$$ (Art. 105)

$\sin 3A = 3\sin A - 4\sin^3 A.$

$\cos 3A = 4\cos^3 A - 3\cos A.$

$$\tan 3A = \frac{3\tan A - \tan^3 A}{1 - 3\tan^2 A}.$$ (Art. 107)

$$\sin\frac{A}{2} = \pm\sqrt{\frac{1 - \cos A}{2}}; \quad \cos\frac{A}{2} = \pm\sqrt{\frac{1 + \cos A}{2}}.$$ (Art. 110)

$$2\sin\frac{A}{2} = \pm\sqrt{1 + \sin A} \pm \sqrt{1 - \sin A}.$$

$$2\cos\frac{A}{2} = \pm\sqrt{1+\sin A} \mp \sqrt{1-\sin A}. \qquad \text{(Art. 113)}$$

$$\tan(A_1 + A_2 + ... + A_n) = \frac{s_1 - s_3 + s_5 - ...}{1 - s_2 + s_4 - ...}. \qquad \text{(Art. 125)}$$

VII. $\log_a mn = \log_a m + \log_a n.$

$\log_a\left(\dfrac{m}{n}\right) = \log_a m - \log_a n.$

$\log_a m^n = n \log_a m.$ $\qquad$ (Art. 136)

$\log_a m = \log_b m \times \log_a b.$ $\qquad$ (Art. 147)

VIII. $\dfrac{\sin A}{a} = \dfrac{\sin B}{b} = \dfrac{\sin C}{c}.$ $\qquad$ (Art. 163.)

$$\cos A = \frac{b^2 + c^2 - a^2}{2bc}, ... \qquad \text{(Art. 164)}$$

$$\sin\frac{A}{2} = \sqrt{\frac{(s-b)(s-c)}{bc}}, ... \qquad \text{(Art. 165)}$$

$$\cos\frac{A}{2} = \sqrt{\frac{s(s-a)}{bc}}, ... \qquad \text{(Art. 166)}$$

$$\tan\frac{A}{2} = \sqrt{\frac{(s-b)(s-c)}{s(s-a)}}, ... \qquad \text{(Art. 167)}$$

$$\sin A = \frac{2}{bc}\sqrt{s(s-a)(s-b)(s-c)}, ... \qquad \text{(Art. 169)}$$

$a = b\cos C + c\cos B, ...$ $\qquad$ (Art. 170)

$$\tan\frac{B-C}{2} = \frac{b-c}{b+c}\cot\frac{A}{2}, ... \qquad \text{(Art. 171)}$$

$$S = \sqrt{s(s-a)(s-b)(s-c)} = \frac{1}{2}bc\sin A = \frac{1}{2}ca\sin B = \frac{1}{2}ab\sin C. \qquad \text{(Art. 198)}$$

IX. $R = \dfrac{a}{2\sin A} = \dfrac{b}{2\sin B} = \dfrac{c}{2\sin C} = \dfrac{abc}{4S}.$ $\qquad$ (Arts. 200, 201)

$r = \dfrac{S}{s} = (s-a)\tan\dfrac{A}{2} = ... = ...$ $\qquad$ (Arts. 202, 203)

$r_1 = \dfrac{S}{s-a} = s\tan\dfrac{A}{2}.$ $\qquad$ (Arts. 205, 206)

Area of a quadrilateral inscribable in a circle

$$= \sqrt{(s-a)(s-b)(s-c)(s-d)}. \qquad \text{(Art. 219)}$$

$\dfrac{\sin\theta}{\theta} = 1$, when θ is very small. (Art. 228)

Area of a circle of radius $r = \pi r^2$. (Art. 233)

X. $\sin\alpha + \sin(\alpha+\beta) + \sin(\alpha+2\beta) + \dots$ to n terms

$$= \dfrac{\sin\left\{\alpha+\dfrac{n-1}{2}\beta\right\}\sin\dfrac{n\beta}{2}}{\sin\dfrac{\beta}{2}}.$$ (Art. 241)

$\cos\alpha + \cos(\alpha+\beta) + \cos(\alpha+2\beta) + \dots$ to n terms

$$= \dfrac{\cos\left\{\alpha+\dfrac{n-1}{2}\beta\right\}\sin\dfrac{n\beta}{2}}{\sin\dfrac{\beta}{2}}.$$ (Art. 242)

PART II

XI. $\underset{n=\infty}{\mathrm{Lt}}\left(1+\dfrac{1}{n}\right)^{n} = e = 2.71828\dots\dots$ (Arts. 2, 3)

$e^{x} = 1 + x + \dfrac{x^2}{\underline{|2}} + \dfrac{x^3}{\underline{|3}} + \dots\dots$ ad inf.

$a^{x} = 1 + x\log_e a + \dfrac{x^2}{\underline{|2}}\left(\log_e a\right)^2 + \dots$ ad inf. (Art. 5)

$\log_e(1+x) = x - \dfrac{1}{2}x^2 + \dfrac{1}{3}x^3 - \dfrac{1}{4}x^4 + \dots\dots$ ad inf.

when $x > -1$ and $\not> 1$. (Art. 8)

$\underset{n=\infty}{\mathrm{Lt}}\left(\cos\dfrac{a}{n}\right)^{n} = \underset{n=\infty}{\mathrm{Lt}}\left(\dfrac{\sin\dfrac{a}{n}}{\dfrac{a}{n}}\right)^{n} = 1.$ (Arts. 14, 15)

XII. $(\cos\theta + i\sin\theta)^{n} = \cos n\theta + i\sin n\theta.$ (Art. 21)

XIII. $\sin n\theta = n\cos^{n-1}\theta\sin\theta$

$\qquad -\dfrac{n(n-1)(n-2)}{1\cdot2\cdot3}\cos^{n-3}\theta\sin^3\theta + \dots\dots$

$\cos n\theta = \cos^{n}\theta - \dfrac{n(n-1)}{1\cdot2}\cos^{n-2}\theta\sin^2\theta$

$$+\frac{n(n-1)(n-2)(n-3)}{1\cdot2\cdot3\cdot4}\cos^{n-4}\theta\sin^{4}\theta-......$$ (Art. 27)

$$\tan n\theta=\frac{s_{1}-s_{3}+s_{5}-s_{7}+......}{1-s_{2}+s_{4}-s_{6}+......}$$ (Art. 30)

XIV. $\sin\alpha=\alpha-\dfrac{\alpha^{3}}{\underline{|3}}+\dfrac{\alpha^{5}}{\underline{|5}}-......$ ad inf. (Art. 33)

$$\cos\alpha=1-\frac{\alpha^{2}}{\underline{|2}}+\frac{\alpha^{4}}{\underline{|4}}-......\text{ ad inf.}$$ (Art. 32)

$$\sin x=\frac{e^{xi}-e^{-xi}}{2i};\quad\cos x=\frac{e^{xi}+e^{-xi}}{2}.$$ (Art. 60)

XV. $\log(\alpha+\beta i)=\log_{e}\sqrt{\alpha^{2}+\beta^{2}}+i(2n\pi+\theta),$

where $\cos\theta=\dfrac{\alpha}{+\sqrt{\alpha^{2}+\beta^{2}}}$ and $\sin\theta=\dfrac{\beta}{+\sqrt{\alpha^{2}+\beta^{2}}}.$ (Art. 82)

XVI. $\tan^{-1}x=x-\dfrac{1}{3}x^{3}+\dfrac{1}{5}x^{5}-\dfrac{1}{7}x^{7}+......$ ad inf.,

where x is numerically not greater than unity. (Art. 95)

$$\theta-p\pi=\tan\theta-\frac{1}{3}\tan^{3}\theta+\frac{1}{5}\tan^{5}\theta-......\text{ ad inf.}$$

where θ lies between $p\pi-\dfrac{\pi}{4}$ and $p\pi+\dfrac{\pi}{4}.$ (Art. 96)

XVII. $x^{2n}-2a^{n}x^{n}\cos n\theta+a^{2n}$

$$=\prod_{r=0}^{r=n-1}\left\{x^{2}-2ax\cos\left(\theta+\frac{2r\pi}{n}\right)+a^{2}\right\}$$ (Art. 115)

$$x^{n}-1=(x^{2}-1)\prod_{r=1}^{r=\frac{n}{2}-1}\left(x^{2}-2x\cos\frac{2r\pi}{n}+1\right),\ (n\text{ even})$$

and $-(x-1)\displaystyle\prod_{r=1}^{r=\frac{n-1}{2}}\left(x^{2}-2x\cos\frac{2r\pi}{n}+1\right),\ (n\text{ odd}).$ (Art. 119)

$$x^{n}+1=\prod_{r=0}^{r=\frac{n}{2}-1}\left(x^{2}-2x\cos\frac{2r+1}{n}\pi+1\right),\ (n\text{ even}).$$

and $\qquad = (x+1) \prod\limits_{r=0}^{r=\frac{n-3}{2}} \left(x^2 - 2x\cos\dfrac{2r+1}{n}\pi + 1 \right), \ (n \text{ odd}).$ (Art. 120)

$$\sin\theta = \theta\left(1 - \frac{\theta^2}{\pi^2}\right)\left(1 - \frac{\theta^2}{2^2\pi^2}\right)\left(1 - \frac{\theta^2}{3^2\pi^2}\right)\ldots\ldots \text{ ad inf.} \qquad \text{(Art. 122)}$$

$$\cos\theta = \left(1 - \frac{4\theta^2}{\pi^2}\right)\left(1 - \frac{4\theta^2}{3^2\pi^2}\right)\left(1 - \frac{4\theta^2}{5^2\pi^2}\right)\ldots\ldots \text{ ad inf.} \qquad \text{(Art. 123)}$$

Exponential and Logarithmic Series

■ **1.** In the following chapter we are about to obtain an expansion in powers of x for the expression a^x, where both a and x are real, and also to obtain an expansion for $\log_e (1 + x)$, where x is real and less than unity, and e stands for a quantity to be defined.

■ **2.** *To find the value of the quantity* $\left(1+\dfrac{1}{n}\right)^n$, *when n becomes infinitely great and is real.*

Since $\dfrac{1}{n} < 1$, we have, by the Binomial Theorem,

$$\left(1+\frac{1}{n}\right)^n = 1 + n\cdot\frac{1}{n} + \frac{n(n-1)}{1\cdot 2}\frac{1}{n^2} + \frac{n(n-1)(n-2)}{1\cdot 2\cdot 3}\frac{1}{n^3} + \ldots$$

$$= 1 + 1 + \frac{1-\dfrac{1}{n}}{1\cdot 2} + \frac{\left(1-\dfrac{1}{n}\right)\left(1-\dfrac{2}{n}\right)}{\underline{3}} + \frac{\left(1-\dfrac{1}{n}\right)\left(1-\dfrac{2}{n}\right)\left(1-\dfrac{3}{n}\right)}{\underline{4}} + \ldots \quad \ldots (1)$$

This series is true for all values of n, however great. Make then n infinite and the right-hand side

$$= 1 + 1 + \frac{1}{\underline{2}} + \frac{1}{\underline{3}} + \frac{1}{\underline{4}} + \ldots \text{ ad inf.} \quad \ldots (2)$$

Hence the limiting value, when n is infinite, of $\left(1+\dfrac{1}{n}\right)^n$ is the sum of the series.

$$1 + 1 + \frac{1}{\underline{2}} + \frac{1}{\underline{3}} + \frac{1}{\underline{4}} \ldots \text{ ad inf.}$$

The sum of this series is always denoted by the quantity e.

Hence we have

$$\underset{n=\infty}{\mathbf{Lt}} \left(1+\frac{1}{\mathbf{n}}\right)^{\mathbf{n}} = \mathbf{e},$$

where $\underset{n=\infty}{\text{Lt}}$ stands for "the limit when $n = \infty$."

Cor. By putting $n = \dfrac{1}{m}$, it follows (since m is zero when n is infinity) that

$$\underset{m=0}{\mathrm{Lt}}\,(1+m)^{1/m} = \underset{n=\infty}{\mathrm{Lt}}\left(1+\frac{1}{n}\right)^{n} = e.$$

■ **3.** This quantity e is finite.

For since

$$\frac{1}{\underline{|3}} < \frac{1}{2\cdot 2} < \frac{1}{2^2},$$

$$\frac{1}{\underline{|4}} < \frac{1}{2\cdot 2\cdot 2} < \frac{1}{2^3},$$

$$\cdots\cdots\cdots\cdots\cdots$$

we have

$$e < 1 + 1 + \frac{1}{2} + \frac{1}{2^2} + \frac{1}{2^3} \ldots\text{ad inf.}$$

$$< 1 + \frac{1}{1 - \dfrac{1}{2}}$$

$$< 1 + 2 \ i.e. \ < 3.$$

Also clearly $e > 2$.

Hence it lies between 2 and 3.

By taking a sufficient number of terms in the series, it can be shown that
$$e = 2.7182818285\ldots$$

■ **4.** *The quantity e is incommensurable.*

For, if possible, suppose it to be equal to a fraction $\dfrac{p}{q}$, where p and q are whole numbers.

We have then

$$\frac{p}{q} = 1 + 1 + \frac{1}{\underline{|2}} + \frac{1}{\underline{|3}} + \ldots + \frac{1}{\underline{|q}} + \frac{1}{\underline{|q+1}} + \frac{1}{\underline{|q+2}} + \ldots \qquad \ldots(1)$$

Multiply this equation by $\underline{|q}$, so that all the terms of the series (1) become integers except those commencing with $\dfrac{\underline{|q}}{\underline{|q+1}}$. Hence we have

$$p\underline{|q} - 1 = \text{whole number} + \frac{\underline{|q}}{\underline{|q+1}} + \frac{\underline{|q}}{\underline{|q+2}} + \frac{\underline{|q}}{\underline{|q+3}} + \ldots,$$

i.e. an integer $= \dfrac{1}{q+1} + \dfrac{1}{(q+1)(q+2)} + \dfrac{1}{(q+1)(q+2)(q+3)} + \ldots \qquad \ldots(2)$

But the right-hand side of this equation is $> \dfrac{1}{q+1}$, and

$$< \frac{1}{q+1} + \frac{1}{(q+1)^2} + \frac{1}{(q+1)^3} + ...,$$

i.e.

$$< \frac{1}{q+1} \div \left(1 - \frac{1}{q+1}\right),$$

i.e.

$$< \frac{1}{q}.$$

Hence the right-hand side of (2) lies between $\dfrac{1}{q+1}$ and $\dfrac{1}{q}$, and therefore a fraction and so cannot be equal to the left-hand side.

Hence our supposition that e was commensurable is incorrect and it therefore must be incommensurable.

■ **5. Exponential Series:** *When x is real, to prove that*

$$e^x = 1 + x + \frac{x^2}{\underline{2}} + \frac{x^3}{\underline{3}} + ...ad\ inf.,$$

and that

$$a^x = 1 + x \log_e a + \frac{x^2}{\underline{2}} (\log_e a)^2 + ...\ ad\ inf.$$

When n is greater than unity, we have

$$\left\{\left(1 + \frac{1}{n}\right)^n\right\}^x = \left(1 + \frac{1}{n}\right)^{nx}$$

$$= 1 + nx \frac{1}{n} + \frac{nx(nx-1)}{1 \cdot 2} \frac{1}{n^2} + \frac{nx(nx-1)(nx-2)}{1 \cdot 2 \cdot 3} \frac{1}{n^3} + ...$$

$$= 1 + x + \frac{x\left(x - \dfrac{1}{n}\right)}{1 \cdot 2} + \frac{x\left(x - \dfrac{1}{n}\right)\left(x - \dfrac{2}{n}\right)}{1 \cdot 2 \cdot 3} + ...$$

In this expression make n infinitely great. The left-hand becomes, as in Art. 2, e^x. The right-hand becomes

$$1 + x + \frac{x^3}{\underline{2}} + \frac{x^3}{\underline{3}} + ...$$

Hence we have

$$e^x = 1 + x + \frac{x^2}{\underline{2}} + \frac{x^3}{\underline{3}} + ...\ ad\ inf. \qquad ...(1)$$

Let $\qquad a = e^c$, so that $c = \log_e a$.

$$\therefore \qquad a^x = e^{cx} = 1 + cx + \frac{c^2 x^2}{\underline{|2}} + \frac{c^3 x^3}{\underline{|3}} + \ldots \text{ ad inf.}$$

by substituting cx for x in the series (1).

$$\therefore \qquad \mathbf{a^x = 1 + x \log_e a + \frac{x^2}{\underline{|2}} (\log_e a)^2 + \frac{x^3}{\underline{|3}} (\log_e a)^3 + \ldots \text{ ad inf.}} \qquad \ldots(2)$$

▮ **6.** It can be shown (as in C. Smith's *Algebra*, Art. 278) that the series (1), and therefore (2), of the last article is convergent for all real values of x.

▮ **7. EXAMPLE 1.** *Prove that* $\frac{1}{2}\left(e - \frac{1}{e}\right) = 1 + \frac{1}{\underline{|3}} + \frac{1}{\underline{|5}} + \ldots$ *ad inf.*

By equation (1) of Art. 5 we have, by putting x in succession equal to 1 and -1,

$$e = 1 + \frac{1}{\underline{|1}} + \frac{1}{\underline{|2}} + \frac{1}{\underline{|3}} + \frac{1}{\underline{|4}} + \ldots \text{ ad inf.}$$

and

$$e^{-1} = 1 - \frac{1}{\underline{|1}} + \frac{1}{\underline{|2}} - \frac{1}{\underline{|3}} + \frac{1}{\underline{|4}} - \ldots \text{ ad inf.}$$

Hence, by subtraction,

$$e - e^{-1} = 2\left(1 + \frac{1}{\underline{|3}} + \frac{1}{\underline{|5}} + \ldots\right),$$

i.e.

$$\frac{1}{2}\left(e - \frac{1}{e}\right) = 1 + \frac{1}{\underline{|3}} + \frac{1}{\underline{|5}} + \ldots \text{ ad inf.}$$

▮ **EXAMPLE 2.** *Find the sum of the series*

$$1 + \frac{1+2}{\underline{|2}} + \frac{1+2+3}{\underline{|3}} + \frac{1+2+3+4}{\underline{|4}} + \ldots \text{ ad inf.}$$

The nth term $= \dfrac{1+2+3+\ldots+n}{\underline{|n}} = \dfrac{\frac{1}{2}n(n+1)}{\underline{|n}}$

$$= \frac{1}{2}\frac{n+1}{\underline{|n-1}} = \frac{1}{2}\left[\frac{(n-1)+2}{\underline{|n-1}}\right] = \frac{1}{2}\left[\frac{1}{\underline{|n-2}} + \frac{2}{\underline{|n-1}}\right],$$

provided that $n > 2$.

Similarly,

the $(n - 1)$th term $= \dfrac{1}{2}\left[\dfrac{1}{\underline{|n-3}} + \dfrac{2}{\underline{|n-2}}\right],$

. . . .

the 4th term $= \dfrac{1}{2}\left[\dfrac{1}{\underline{|2}} + \dfrac{2}{\underline{|3}}\right],$

the 3rd term $= \dfrac{1}{2}\left[\dfrac{1}{\underline{1}} + \dfrac{2}{\underline{2}}\right].$

Also   the 2nd term $= \dfrac{1}{2}\left[1 + \dfrac{2}{\underline{1}}\right].$

and   the 1st term $= \dfrac{1}{2}\left[\dfrac{2}{1}\right].$

Hence, by addition, the whole series

$$= \frac{1}{2}\left[1 + \frac{1}{\underline{1}} + \frac{1}{\underline{2}} + \frac{1}{\underline{3}} + \dots \text{ ad inf.}\right]$$

$$+ \frac{1}{2}\cdot 2\left[1 + \frac{1}{\underline{1}} + \frac{1}{\underline{2}} + \frac{1}{\underline{3}} + \dots \text{ ad inf.}\right]$$

$$= \frac{1}{2}\cdot e + e = \frac{3e}{2}.$$

■ **8. Logarithmic Series:** *To prove that, when y is real and numerically < 1, then*

$$\log_e(1 + y) = y - \frac{1}{2}y^2 + \frac{1}{3}y^3 - \frac{1}{4}y^4 + \dots \text{ ad inf.}$$

In the equation (2) of Art. 5, put
$$a = 1 + y,$$
and we have

$$(1 + y)^x = 1 + x\log_e(1 + y) + \frac{x^2}{\underline{2}}\left\{\log_e(1 + y)\right\}^2 + \dots \qquad \dots(1)$$

But, since y is real and numerically < unity, we have

$$(1 + y)^x = 1 + x\cdot y + \frac{x(x-1)}{1\cdot 2}y^2 + \frac{x(x-1)(x-2)}{1\cdot 2\cdot 3}y^3 + \dots \qquad \dots(2)$$

The series on the right-hand side of (1) and (2) are equal to one another and both convergent, when y is numerically < 1. Also it could be shown that the series on the right-hand side of (2) is convergent when it is arranged in powers of x. Hence we may equate like powers of x.

Thus we have

$$\log_e(1 + y) = y - \frac{y^2}{1\cdot 2} + \frac{(-1)(-2)}{1\cdot 2\cdot 3}y^3 + \frac{(-1)(-2)(-3)}{1\cdot 2\cdot 3\cdot 4}y^4 + \dots \text{ ad inf.},$$

i.e. $$\mathbf{\log_e(1 + y) = y - \frac{1}{2}y^2 + \frac{1}{3}y^3 - \frac{1}{4}y^4 + \text{ ad inf.}} \qquad \dots(3)$$

■ **9.** If $y = 1$, the series (3) of the previous article is equal to

$$1 - \frac{1}{2} + \frac{1}{3} - \frac{1}{4} + \dots \text{ ad inf.}$$

which is known to be convergent.

If $y = -1$, it equals $-1 - \frac{1}{2} - \frac{1}{3} - \frac{1}{4} \dots$ ad inf. which is known to be divergent.

In addition therefore to being true for all values of y between -1 and $+1$, it is true for the value $y = 1$; it is not however true for the value $y = -1$.

■ **10. Calculation of logarithms to base e.**

In the logarithmic series, if we put $y = 1$, we have

$$\log_e 2 = 1 - \frac{1}{2} + \frac{1}{3} - \frac{1}{4} + \dots \text{ ad inf.} \qquad \dots(1)$$

If we put $\qquad y = \frac{1}{2},$

we have

$$\log_e 3 - \log_e 2 = \log_e \frac{3}{2} = \log_e \left(1 + \frac{1}{2}\right)$$

$$= \frac{1}{2} - \frac{1}{2} \cdot \frac{1}{2^2} + \frac{1}{3} \cdot \frac{1}{2^3} - \frac{1}{4} \cdot \frac{1}{2^4} + \dots \qquad \dots(2)$$

If we put $\qquad y = \frac{1}{3}.$

we have

$$\log_e 4 - \log_e 3 = \log_e \left(1 + \frac{1}{3}\right) = \frac{1}{3} - \frac{1}{2} \cdot \frac{1}{3^2} + \frac{1}{3} \cdot \frac{1}{3^3} - \frac{1}{4} \cdot \frac{1}{3^4} + \dots \qquad \dots(3)$$

From these equations we could, by taking a sufficient number of terms, calculate $\log_e 2$, $\log_e 3$ and $\log_e 4$.

It would be found that a large number of terms would have to be taken to give the values of these logarithms to the required degree of accuracy. We shall therefore obtain more convenient series.

■ **11.** By Art. 8 we have

$$\log_e (1+y) y - \frac{1}{2} y^2 + \frac{1}{3} y^3 - \frac{1}{4} y^4 + \dots \qquad \dots(1)$$

and, by changing the sign of y,

$$\log_e (1-y) = -y - \frac{1}{2} y^2 - \frac{1}{3} y^3 - \frac{1}{4} y^4 + \dots \qquad \dots(2)$$

In order that both these series may be true y must be numerically less than unity.

By subtraction, we have

$$\log_e \left(1+y\right) - \log_e \left(1-y\right) = \log_e \frac{1+y}{1-y} = 2\left[y + \frac{1}{3}y^3 + \frac{1}{5}y^5 + ...\right] \qquad ...(3)$$

Let
$$y = \frac{m-n}{m+n},$$

where m and n are positive integers and $m > n$, so that

$$\frac{1+y}{1-y} = \frac{m}{n}.$$

The equation (3) becomes

$$\log_e \frac{m}{n} = 2\left[\left(\frac{m-n}{m+n}\right) + \frac{1}{3}\left(\frac{m-n}{m+n}\right)^3 + \frac{1}{5}\left(\frac{m-n}{m+n}\right)^5 + ...\right] \qquad ...(4)$$

Put $m = 2$, $n = 1$ in (4) and we get $\log_e 2$.

Put $m = 3$, $n = 2$ and we get $\log_e 3 - \log_e 2$, and therefore $\log_e 3$.

By proceeding in this way we get the value of the logarithm of any number to base e.

▇ **12. Logarithms to base 10.** The logarithms of the previous article, to base e, are called Napierian or natural logarithms.

We can convert these logarithms into logarithms to base 10.

For, by Art. 147 (Part I.), we have, if N be any number,

$$\log_e N = \log_{10} N \times \log_e 10.$$

$$\therefore \log_{10} N = \log_e N \times \frac{1}{\log_e 10}.$$

Now, $\log_e 10$ can be found as in the last article and then $\dfrac{1}{\log_e 10}$ is found to be 0.4342944819...,

Hence, $\qquad \log_{10} N = \log_e N \times 0.43429448...,$

so that the logarithm of any number to base 10 is found by multiplying its logarithm to base e by the quantity 0.43429448... This quantity is called the Modulus.

EXAMPLES I

Prove that

1. $\dfrac{1}{2}\left(e + e^{-1}\right) = 1 + \dfrac{1}{\lfloor 2} + \dfrac{1}{\lfloor 4} + \dfrac{1}{\lfloor 6} + ...$

2. $\left(1 + \dfrac{1}{\lfloor 1} + \dfrac{1}{\lfloor 2} + \dfrac{1}{\lfloor 3} + ...\right)\left(1 - \dfrac{1}{\lfloor 1} + \dfrac{1}{\lfloor 2} - \dfrac{1}{\lfloor 3} + ...\right) = 1.$

3. $\left(1+\dfrac{1}{\lfloor 2}+\dfrac{1}{\lfloor 4}+\dfrac{1}{\lfloor 6}+...\right)^2 = 1+\left(1+\dfrac{1}{\lfloor 3}+\dfrac{1}{\lfloor 5}+...\right)^2 ...$

4. $1+\dfrac{2}{\lfloor 3}+\dfrac{3}{\lfloor 5}+\dfrac{4}{\lfloor 7}+... = \dfrac{e}{2}.$

5. $\dfrac{2}{\lfloor 3}+\dfrac{4}{\lfloor 5}+\dfrac{6}{\lfloor 7}+... = e^{-1}.$

6. $\dfrac{\dfrac{1}{\lfloor 2}+\dfrac{1}{\lfloor 4}+\dfrac{1}{\lfloor 6}+...}{1+\dfrac{1}{\lfloor 3}+\dfrac{1}{\lfloor 5}+...} = \dfrac{e-1}{e+1}.$

7. $1+\dfrac{2^3}{\lfloor 2}+\dfrac{3^3}{\lfloor 3}+\dfrac{4^3}{\lfloor 4}+... = 5e.$

Find the sum of the series

8. $1-\dfrac{1}{2}+\dfrac{1}{3}-\dfrac{1}{4}+...$ ad inf.

9. $\dfrac{1}{2}-\dfrac{1}{2}\cdot\dfrac{1}{2^2}+\dfrac{1}{3}\cdot\dfrac{1}{2^3}-\dfrac{1}{4}\cdot\dfrac{1}{2^4}+...$ ad inf.

Prove that

10. $\dfrac{a-b}{a}+\dfrac{1}{2}\left(\dfrac{a-b}{a}\right)^2+\dfrac{1}{3}\left(\dfrac{a-b}{a}\right)^3+... = \log_e a = -\log_e b.$

11. $\log_e \dfrac{1+x}{1-x} = 2\left(x+\dfrac{1}{3}x^3+\dfrac{1}{5}x^5+...\text{ad inf.}\right).$

12. $\log_e \dfrac{x+1}{x-1} = 2\left(\dfrac{1}{x}+\dfrac{1}{3x^3}+\dfrac{1}{5x^5}+...\text{ad inf.}\right),$ if $x>1.$

13. $\log_e (1+3x+2x^2) = 3x-\dfrac{5x^2}{2}+\dfrac{9x^3}{3}-\dfrac{17x^4}{4}+...+(-1)^{n-1}\dfrac{2^n+1}{n}x^n+...,$

provided that $2x$ be not $>1.$

14. $2\log_e x - \log_e(x+1) - \log_e(x-1) = \dfrac{1}{x^2}+\dfrac{1}{2x^4}+\dfrac{1}{3x^6}+...,$ if $x>1.$

15. $\log_e 2 = \dfrac{1}{1\cdot 2}+\dfrac{1}{3\cdot 4}+\dfrac{1}{5\cdot 6}+...$ ... ad inf.

16. $\log_e 2 - \dfrac{1}{2} = \dfrac{1}{1\cdot 2\cdot 3}+\dfrac{1}{3\cdot 4\cdot 5}+\dfrac{1}{5\cdot 6\cdot 7}+...$ ... ad inf

17. $\tan\theta+\dfrac{1}{3}\tan^3\theta+\dfrac{1}{5}\tan^5\theta+... = \dfrac{1}{2}\log\dfrac{\cos\left(\theta-\dfrac{\pi}{4}\right)}{\cos\left(\theta+\dfrac{\pi}{4}\right)},$ if $\theta<\dfrac{\pi}{4}.$

18. If θ be $>\dfrac{\pi}{2}$ and $<\pi$, prove that

$(1)\sin\theta+\dfrac{1}{3}\sin^3\theta+\dfrac{1}{5}\sin^5\theta+......$ ad inf.

$= 2\left[\cot\dfrac{\theta}{2}+\dfrac{1}{3}\cot^3\dfrac{\theta}{2}+\dfrac{1}{5}\cot^5\dfrac{\theta}{2}+...\text{ad inf.}\right],$

and, if θ be > 0 and $< \dfrac{\pi}{2}$, prove that

$$(2)\frac{1}{2}\sin^2\theta + \frac{1}{4}\sin^4\theta + \frac{1}{6}\sin^6\theta + \dots \dots \text{ad inf.}$$

$$= 2\left[\tan^2\frac{\theta}{2} + \frac{1}{3}\tan^6\frac{\theta}{2} + \frac{1}{5}\tan^{10}\frac{\theta}{2} + \dots \text{ad inf.}\right]$$

19. If $\tan^2\theta < 1$, prove that

$$\tan^2\theta - \frac{1}{2}\tan^4\theta + \frac{1}{3}\tan^6\theta - \dots \text{ad inf.}$$

$$= \sin^2\theta + \frac{1}{2}\sin^4\theta + \frac{1}{3}\sin^6\theta + \dots \text{ad inf.}$$

20. Prove that, if 2θ be not a multiple of π,

$$\log\cot\theta = \cos 2\theta + \frac{1}{3}\cos^3 2\theta + \frac{1}{5}\cos^5 2\theta + \dots \text{ad inf.}$$

21. Prove that the coefficient of x^n in the expansion of
$$\{\log_e(1+x)\}^2$$

is $\qquad \dfrac{2(-1)^n}{n}\left[1 + \frac{1}{2} + \frac{1}{3} + \dots + \frac{1}{n-1}\right].$

22. Use the methods of Arts. 11 and 12 to prove that
$$\log_{10} 2 = 0.30103\dots$$
and $\qquad\qquad\qquad \log_{10} 3 = 0.47712\dots$

23. Draw the curve $y = \log_e x$.
    [If x be negative, y is imaginary; when x is zero, y equals $-\infty$; when x is the unity, y is nothing; when x is positive and > 1, y is always positive; when x is infinity, y is infinity also.]

24. Draw the curve $y = \log_{10} x$ and state the geometrical relation between it and the curve of the last example.
    [Use Art. 147, Part I.]

25. Draw the curve $y = a^x$.

■ **13.** The two following limits will be required in the next chapter but one.

■ **14.** *To prove that the value of* $\left(\cos\dfrac{a}{n}\right)^n$, *when n is infinite, is unity.*

We have $\qquad \cos\dfrac{a}{n} = \left(1 - \sin^2\dfrac{a}{n}\right)^{\frac{1}{2}}.$

$\therefore \qquad \left(\cos\dfrac{a}{n}\right)^n = \left(1 - \sin^2\dfrac{a}{n}\right)^{\frac{n}{2}} = \left[\left(1 - \sin^2\dfrac{a}{n}\right)^{-\frac{1}{\sin^2\frac{a}{n}}}\right]^{-\frac{n}{2}\sin^2\frac{a}{n}}.$

Now, by putting

$$-\sin^2 \frac{a}{n} = m,$$

We have

$$\text{Lt}_{n=\infty} \left\{1 - \sin^2 \frac{a}{n}\right\}^{-\frac{1}{\sin^2 \frac{a}{n}}} = \text{Lt}_{m=0} \left\{1 - m\right\}^{\frac{1}{m}} = e. \qquad \text{(Art. 2, Cor.)}$$

Also, by Art. 228 (Part I.),

$$\frac{n}{2} \sin^2 \frac{a}{n}$$

$$= \left(\frac{\sin \frac{a}{n}}{\frac{a}{n}}\right)^2 \times \frac{a^2}{2n} = 1 \times 0 = 0,$$

when n is infinite.

Hence, when n is infinite,

$$\left[\cos \frac{a}{n}\right]^n = e^0 = 1.$$

Alter. This limit may also be found by using the logarithmic series.

For, putting $\left(\cos \frac{a}{n}\right)^n = u$, we have

$$\log_e u = n \log_e \cos \frac{a}{n} = \frac{n}{2} \log_e \cos^2 \frac{a}{n}$$

$$= \frac{n}{2} \log_e \left(1 - \sin^2 \frac{a}{n}\right)$$

$$= -\frac{n}{2} \left(\sin^2 \frac{a}{n} + \frac{1}{2} \sin^4 \frac{a}{n} + \frac{1}{3} \sin^6 \frac{a}{n} + \dots\right).$$

$$\text{(Art. 8.)}$$

The series inside the bracket lies between $\sin^2 \frac{a}{n}$ and the series

$$\sin^2 \frac{a}{n} + \sin^4 \frac{a}{n} + \sin^6 \frac{a}{n} + \dots \text{ ad inf.,}$$

i.e. lies between

$$\sin^2 \frac{a}{n} \text{ and } \frac{\sin^2 \frac{a}{n}}{1 - \sin^2 \frac{a}{n}},$$

i.e. lies between $\sin^2 \dfrac{a}{n}$ and $\tan^2 \dfrac{a}{n}$.

hence $-\log u$ lies between

$$\frac{n}{2}\sin^2 \frac{a}{n} \text{ and } \frac{n}{2}\tan^2 \frac{a}{n} \qquad \qquad \text{...(1)}$$

But

$$\operatorname*{Lt}_{n=\infty} \frac{n}{2}\sin^2 \frac{a}{n} = \operatorname*{Lt}_{n=\infty} \left(\frac{\sin \dfrac{a}{n}}{\dfrac{a}{n}} \right)^2 \times \frac{a^2}{2n} = 1 \times 0 = 0,$$

and

$$\operatorname*{Lt}_{n=\infty} \frac{n}{2}\tan^2 \frac{a}{n} = \operatorname*{Lt}_{n=\infty} \left\{ \left(\frac{\sin \dfrac{a}{n}}{\dfrac{a}{n}} \right)^2 \times \frac{1}{\cos^2 \dfrac{a}{n}} \times \frac{a^2}{2n} \right\} = 1 \times 1 \times 0 = 0.$$

(Art. 228, Part I.)

Hence in the limit both quantities (1) become 0, so that $\log u$, becomes zero also, and therefore, in the limit,

$$u = 1.$$

▮ **15.** *To prove that the limiting value of* $\left(\dfrac{\sin \dfrac{a}{n}}{\dfrac{a}{n}} \right)^n$, *when n is infinite, is unity.*

We have shown, in Art. 227 (Part I.), that $\sin \theta$, θ and $\tan \theta$ are in the ascending order of magnitude.

Hence $\sin \dfrac{a}{n}, \dfrac{a}{n}$ and $\tan \dfrac{a}{n}$

are in the ascending order.

Hence $\qquad 1, \dfrac{\dfrac{a}{n}}{\sin \dfrac{a}{n}}, \text{ and } \dfrac{1}{\cos \dfrac{a}{n}}$

are in ascending order.

Therefore $\left(\dfrac{\dfrac{a}{n}}{\sin \dfrac{a}{n}} \right)^n$ lies between 1 and $\left(\dfrac{1}{\cos \dfrac{a}{n}} \right)^n$, so that $\left(\dfrac{\sin \dfrac{a}{n}}{\dfrac{a}{n}} \right)$ lies between

1 and $\left(\cos \dfrac{a}{n} \right)^n$.

But, by the last article, the value of $\left(\cos\dfrac{a}{n}\right)^n$ is unity,

when n is infinite

Hence, when n is infinite, the value of $\left(\dfrac{\sin\dfrac{a}{n}}{\dfrac{a}{n}}\right)^n$ is unity.

■ **16.** There is one point in Art. 2 that requires some examination.

We ought to show rigidly that the value of the series on the right hand of (1) is equal, when n becomes indefinitely great, to the series (2).

Take the $(p + 1)$th term of the series (1), viz.

$$\frac{\left(1-\dfrac{1}{n}\right)\left(1-\dfrac{2}{n}\right)......\left(1-\dfrac{p-1}{n}\right)}{\lfloor p}\qquad ...(1)$$

When a, b, c ... are all positive quantities and less than unity, we have
$$(1 - a)(1 - b) = 1 - a - b + ab > 1 - a - b,$$
and $\quad (1 - a)(1 - b)(1 - c) > (1 - a - b)(1 - c) > 1 - a - b - c,$
and so on, so that
$$(1 - a)(1 - b)(1 - c) .... > 1 - (a + b + c + ...).$$
Hence the numerator of (1) lies between unity and

$$1-\left(\frac{1}{n}+\frac{2}{n}+\frac{3}{n}+......+\frac{p-1}{n}\right),$$

i.e. between unity and $1-\dfrac{p(p-1)}{2n}$.

Therefore the quantity (1) lies between

$$\frac{1}{\lfloor p}\quad\text{and}\quad\frac{1}{\lfloor p}-\frac{1}{2n}\frac{1}{\lfloor p-2}.$$

Hence the whole series (1) of Art. 2 lies between

$$1+1+\frac{1}{\lfloor 2}+\frac{1}{\lfloor 3}+......\text{ad inf.,}$$

and $\qquad 1+1+\left(\dfrac{1}{\lfloor 2}-\dfrac{1}{2n}\right)+\left(\dfrac{1}{\lfloor 3}-\dfrac{1}{2n}\right)+\left(\dfrac{1}{\lfloor 4}-\dfrac{1}{2n}\dfrac{1}{\lfloor 2}\right)+...\text{ ad inf.,}$

i.e. $\qquad 1+1+\dfrac{1}{\lfloor 2}+\dfrac{1}{\lfloor 3}+...\text{ ad inf.}$

$$-\frac{1}{2n}\left(1+\frac{1}{\lfloor 2}+\frac{1}{\lfloor 3}+...\text{ ad inf.}\right).$$

Now the series $1 + \dfrac{1}{\lfloor 2} + \dfrac{1}{\lfloor 3} + \ldots$ ad inf. is, as in Art. 6, convergent, so that the quantity

$\dfrac{1}{2n}\left(1 + \dfrac{1}{\lfloor 2} + \dfrac{1}{\lfloor 3} + \ldots\right)$ is, when x is made indefinitely great, ultimately equal to zero.

Therefore, finally, the series (1) of Art. 2 is equal, in the limit, to

$$1 + 1 + \dfrac{1}{\lfloor 2} + \dfrac{1}{\lfloor 3} + \dfrac{1}{\lfloor 4} + \ldots \text{ ad inf.}$$

A similar argument will apply to the series in Art. 5 and also to those in Arts. 32 and 33.

CHAPTER 2

Complex Quantities

■ **17. Complex quantities:** The quantity $x + y\sqrt{-1}$, where x and y are both real, is called a complex quantity. A complex quantity consists therefore of the sum of two quantities, one of which is wholly real and the other of which is wholly imaginary.

■ **18.** A complex quantity can always be put into the form $r(\cos\theta + \sqrt{-1}\sin\theta)$, where r and θ are both real.

For assume that

$$x + y\sqrt{-1} = r(\cos\theta + \sqrt{-1}\sin\theta)$$

$$= r\cos\theta + \sqrt{-1}\ r\sin\theta.$$

Equating the real and imaginary parts on the two sides of this equation, we have

$$r\cos\theta = x \qquad \qquad \text{...(1)}$$

and $\qquad\qquad r\sin\theta = y \qquad\qquad\qquad\qquad\qquad$ ...(2)

Hence, by squaring and adding, we have $r^2 = x^2 + y^2$,

so that $\qquad\qquad\qquad r = \sqrt{x^2 + y^2}.$

It is customary to take the positive square root of $x^2 + y^2$ and hence r is known. From (1) and (2) we then have

$$\cos\theta = \frac{x}{\sqrt{x^2 + y^2}} \text{ and } \sin\theta = \frac{y}{\sqrt{x^2 + y^2}}.$$

Whatever be the values of x and y, there is one value of θ, and only one value, lying between $-\pi$ radian and $+\pi$ radians which satisfies these two equations.

The quantity $x + y\sqrt{-1}$ can therefore always be expressed in the form $r(\cos\theta + \sqrt{-1}\sin\theta)$.

Def. The quantity $+\sqrt{x^2 + y^2}$ is called the **Modulus** of the complex quantity, and that value of θ (lying between $-\pi$ and $+\pi$) which satisfies the relations

$$\cos\theta = \frac{x}{+\sqrt{x^2 + y^2}} \text{ and } \sin\theta = \frac{y}{+\sqrt{x^2 + y^2}}$$

is called the principal value of the **Amplitude** of

$$x + y\sqrt{-1}.$$

■ **19. EXAMPLE 1.** *Express in the above form the quantity* $1 + \sqrt{-1}.$

Here
$$1 + \sqrt{-1} = r(\cos\theta + \sqrt{-1}\sin\theta),$$

so that
$$r\cos\theta = 1,$$

and
$$r\sin\theta = 1.$$

We therefore have
$$r = +\sqrt{1+1} = +\sqrt{2},$$

and then
$$\cos\theta = \frac{1}{\sqrt{2}} \text{ and } \sin\theta = \frac{1}{\sqrt{2}},$$

so that
$$\theta = \frac{\pi}{4}.$$

Hence
$$1 + \sqrt{-1} = \sqrt{2}\left[\cos\frac{\pi}{4} + \sqrt{-1}\sin\frac{\pi}{4}\right],$$

so that $\sqrt{2}$ is the modulus and $\frac{\pi}{4}$ is the principal value of the amplitude of the given expression.

■ **EXAMPLE 2.** *Quantity* $-1 + \sqrt{-3}.$

Here
$$-1 + \sqrt{-1}\sqrt{3} = r(\cos\theta + \sqrt{-1}\sin\theta),$$

so that
$$r\cos\theta = -1, \text{ and } r\sin\theta = \sqrt{3}.$$

∴
$$r = +\sqrt{1+3} = +2,$$

and then,
$$\cos\theta = -\frac{1}{2} \text{ and } \sin\theta = \frac{\sqrt{3}}{2},$$

so that
$$\theta = \frac{2\pi}{3}.$$

∴
$$-1 + \sqrt{-3} = 2\left[\cos\frac{2\pi}{3} + \sqrt{-1}\sin\frac{2\pi}{3}\right].$$

■ **EXAMPLE 3.** *Quantity* $-1 - \sqrt{-3}.$

Here $r\cos\theta = -1$, and $r\sin\theta = -\sqrt{3}$,

so that $r = +\sqrt{1+3} = +2$, $\cos\theta = -\frac{1}{2}$ and $\sin\theta = -\frac{\sqrt{3}}{2}$.

Hence (since we choose for θ that value which lies between $-\pi$ and $+\pi$) we have

$$\theta = -\frac{2\pi}{3}.$$

∴
$$-1 - \sqrt{-3} = 2\left[\cos\left(-\frac{2\pi}{3}\right) + i\sin\left(-\frac{2\pi}{3}\right)\right].$$

■ **20.** In Art 18 the equations

$$\cos\theta = \frac{x}{+\sqrt{x^2+y^2}} \quad \text{and} \quad \sin\theta = \frac{y}{+\sqrt{x^2+y^2}}$$

are satisfied by more than one value of θ. For the cosine and sine of an angle repeat the same values when the angle is increased by any multiple of 2π radians, so that, if θ denote the value between $-\pi$ and $+\pi$ satisfying the above relations, the general solution is

$$2n\pi + \theta,$$

where n is any integer.

This is expressed by saying that the amplitude of a complex quantity is **many-valued.** The principal value is that particular value of the amplitude that lies between $-\pi$ and $+\pi$.

If to the principal value of θ we add any multiple of 2π we obtain one of its many values.

To sum up: If θ be that value, lying between $-\pi$ and $+\pi$, which satisfies the equations

$$\cos\theta = \frac{x}{\sqrt{x^2+y^2}} \quad \text{and} \quad \sin\theta = \frac{y}{\sqrt{x^2+y^2}} \qquad ...(1)$$

then

$$x + y\sqrt{-1} = \sqrt{x^2+y^2}\left[\cos(2n\pi+\theta)+\sqrt{-1}\sin(2n\pi+\theta)\right].$$

The quantity $2n\pi + \theta$ is called the amplitude and θ is called its principal value. For brevity we often write equations (1) in the form

$$\tan\theta = \frac{y}{x}, \quad i.e. \quad \theta = \tan^{-1}\frac{y}{x},$$

but it must be understood that here the angle denoted is the one that satisfies the conditions (1).

■ **21. De Moivre's Theorem.** *Whatever may be the value of n, positive or negative, integral or fractional, the value or* **one** *of the values, of*

$$(\cos\theta + \sqrt{-1}\sin\theta)^n \quad \text{is} \quad \cos n\theta + \sqrt{-1}\sin n\theta:$$

Case I. Let n be a positive integer.

By simple multiplication we have

$$[\cos\alpha + \sqrt{-1}\sin\alpha]\,[\cos\beta + \sqrt{-1}\sin\beta]$$

$$= \cos\alpha\cos\beta - \sin\alpha\sin\beta + \sqrt{-1}\,[\sin\alpha\cos\beta + \cos\alpha\sin\beta]$$

$$= \cos(\alpha+\beta) + \sqrt{-1}\sin(\alpha+\beta).$$

So $\quad [\cos\alpha + \sqrt{-1}\sin\alpha][\cos\beta + \sqrt{-1}\sin\beta][\cos\gamma + \sqrt{-1}\sin\gamma]$

$$= [\cos(\alpha+\beta) + \sqrt{-1}\,\sin(\alpha+\beta)][\cos\gamma + \sqrt{-1}\,\sin\gamma]$$

$$= [\cos(\alpha+\beta)\cos\gamma - \sin(\alpha+\beta)\sin\gamma]$$

$$+ \sqrt{-1}\,[\sin(\alpha+\beta)\cos\gamma + \cos(\alpha+\beta)\sin\gamma]$$

$$= \cos(\alpha+\beta+\gamma) + \sqrt{-1}\,\sin(\alpha+\beta+\gamma).$$

This process may evidently be continued indefinitely, so that

$$[\cos \alpha + \sqrt{-1} \sin \alpha][\cos \beta + \sqrt{-1} \sin \beta][\cos \gamma + \sqrt{-1} \sin \gamma] \quad \text{... to } n \text{ factors}$$

$$= \cos (\alpha + \beta + \gamma + \text{... to } n \text{ terms}) + \sqrt{-1} \sin [\alpha + \beta + \gamma + \text{... to } n \text{ terms}].$$

In this expression put

$$\alpha = \beta = \gamma = \ldots\ldots = \theta,$$

so that we have

$$[\cos \theta + \sqrt{-1} \sin \theta]^n = \cos n\theta + \sqrt{-1} \sin n\theta.$$

Case II. Let n be a negative integer and equal to $-m$.

We have, by the ordinary law of indices,

$$(\cos \theta + \sqrt{-1} \sin \theta)^n = (\cos \theta + \sqrt{-1} \sin \theta)^{-m}$$

$$= \frac{1}{(\cos\theta + \sqrt{-1}\sin\theta)^m} = \frac{1}{\cos m\theta + \sqrt{-1}\sin m\theta}, \quad \text{by Case I,}$$

$$= \frac{\cos m\theta - \sqrt{-1}\sin m\theta}{(\cos m\theta + \sqrt{-1}\sin m\theta)(\cos m\theta - \sqrt{-1}\sin m\theta)}$$

$$= \frac{\cos m\theta - \sqrt{-1}\sin m\theta}{\cos^2 m\theta + \sin^2 m\theta} = \cos m\theta - \sqrt{-1}\sin m\theta$$

$$= \cos (-m)\theta + \sqrt{-1} \sin (-m)\theta$$

$$= \cos n\theta + \sqrt{-1} \sin n\theta.$$

Case III. Let n be fractional and equal to $\dfrac{p}{q}$, where q is a positive integer and p is an integer, positive or negative.

By the previous cases, we have

$$\left[\cos\frac{\theta}{q} + \sqrt{-1}\sin\frac{\theta}{q}\right]^q = \cos\left(q \cdot \frac{\theta}{q}\right) + \sqrt{-1}\sin\left(q \cdot \frac{\theta}{q}\right)$$

$$= \cos \theta + \sqrt{-1} \sin \theta.$$

Therefore $\cos\dfrac{\theta}{q} + \sqrt{-1}\sin\dfrac{\theta}{q}$ is such that its qth power is $\cos \theta + \sqrt{-1} \sin \theta$.

Hence $\cos\dfrac{\theta}{q} + \sqrt{-1}\sin\dfrac{\theta}{q}$ is one of the qth roots of

$$\cos \theta + \sqrt{-1} \sin \theta,$$

i.e.
$$\cos\frac{\theta}{q} + \sqrt{-1}\sin\frac{\theta}{q}$$

is one of the values of

$$\left(\cos \theta + \sqrt{-1} \sin \theta\right)^{1/q}$$

Raise each of these quantities to the pth power.

We then have that one of the values of

$$\left[\cos \theta + \sqrt{-1}\sin \theta\right]^{p/q} \text{ is } \left(\cos\frac{\theta}{q} + \sqrt{-1}\sin\frac{\theta}{q}\right)^{p},$$

i.e.
$$\cos\frac{p\theta}{q} + \sqrt{-1}\sin\frac{p\theta}{q}.$$

■ **22.** The quantity i is always used to denote $\sqrt{-1}$ and will be often so used hereafter. The expression $\cos \theta + i \sin \theta$ therefore means $\cos \theta + \sqrt{-1}\sin \theta$.

■ **EXAMPLE 1.** *Simplify*

$$\frac{\left(\cos 3\theta + i\sin 3\theta\right)^5 \left(\cos \theta - i\sin \theta\right)^3}{\left(\cos 5\theta + i\sin 5\theta\right)^7 \left(\cos 2\theta - i\sin 2\theta\right)^5}$$

We have　　　$\cos 3\theta + i \sin 3\theta = (\cos \theta + i \sin \theta)^3$,

$\cos \theta - i \sin \theta = \cos(-\theta) + i \sin(-\theta) = (\cos \theta + i \sin \theta)^{-1}$,

$\cos 5\theta + i \sin 5\theta = (\cos \theta + i \sin \theta)^5$,

and　　　$\cos 2\theta - i \sin 2\theta = \cos(-2\theta) + i \sin(-2\theta) = (\cos \theta + i \sin \theta)^{-2}$.

The given expression therefore

$$= \frac{\left(\cos \theta + i\sin \theta\right)^{15} \left(\cos \theta + i\sin \theta\right)^{-3}}{\left(\cos \theta + i\sin \theta\right)^{35} \left(\cos \theta + i\sin \theta\right)^{-10}}$$

$$= (\cos \theta + i \sin \theta)^{-13} = \cos 13\theta - i \sin 13\theta$$

■ **EXAMPLE 2.** *If* $2 \cos \theta = x + \dfrac{1}{x}$ *and* $2 \cos \phi = y + \dfrac{1}{y}$, *prove that one of the values*

of $x^m y^n + \dfrac{1}{x^m y^n}$ *is* $2 \cos(m\theta + n\phi)$.

We have　　　$x^2 - 2x \cos \theta = -1$.

∴　　　$(x - \cos \theta)^2 = -1 + \cos^2 \theta = -\sin^2 \theta$.

∴　　　$x = \cos \theta + i \sin \theta$,

so that　　　$x^m = \cos m\theta + i \sin m\theta$,

and　　　$\dfrac{1}{x^m} = \cos m\theta - i \sin m\theta$.

Similarly　　　$y = \cos \phi + i \sin \phi$,

so that　　　$y^n = \cos n\phi + i \sin n\phi$,

and　　　$\dfrac{1}{y^n} = \cos n\phi - i \sin n\phi$.

∴　　　$x^m y^n + \dfrac{1}{x^m y^n}$

$$= (\cos m\theta + i \sin m\theta)(\cos n\phi + i \sin n\phi)$$
$$+ (\cos m\theta - i \sin m\theta)(\cos m\phi - i \sin n\phi)$$
$$= \cos(m\theta + n\phi) + i \sin(m\theta + n\phi) + \cos(m\theta + n\phi)$$
$$- i \sin(m\theta + n\phi)$$
$$= 2\cos(m\theta + n\phi).$$

Similarly, it could be shown that one of the values of

$$\frac{x^m}{y^n} + \frac{y^n}{x^m} \text{ is } 2\cos(m\theta - n\phi).$$

▮ **EXAMPLE 3.** *If* $\sin \alpha + \sin \beta + \sin \gamma = \cos \alpha + \cos \beta + \cos \gamma = 0$,
prove that $\cos 3\alpha + \cos 3\beta + \cos 3\gamma = 3 \cos(\alpha + \beta + \gamma)$,
and $\sin 3\alpha + \sin 3\beta + \sin 3\gamma = 3 \sin(\alpha + \beta + \gamma)$.

This is an example of the many trigonometrical identities which are derived from algebraical identities.

For we know that if $a + b + c = 0$,

then, $a^2 + b^2 + c^2 = 3abc$.

Let $a = \cos a + i \sin a$, $b = \cos \beta + i \sin \beta$, and $c = \cos \gamma + i \sin \gamma$,
so that we have $a + b + c = 0$.

$\therefore$ $(\cos \alpha + i \sin \alpha)^2 + (\cos \beta + i \sin \beta)^2 + (\cos \gamma + i \sin \gamma)^2$
$$= 3(\cos \alpha + i \sin \alpha)(\cos \beta + i \sin \beta)(\cos \gamma + i \sin \gamma),$$
so that, by De Moivre's Theorem,

$$(\cos 3\alpha + \cos 3\beta + \cos 3\gamma) + i(\sin 3\alpha + \sin 3\beta + \sin 3\gamma)$$
$$= 3 \cos(\alpha + \beta + \gamma) + 3i \sin(\alpha + \beta + \gamma).$$

Hence, by equating real and imaginary parts, we have the required results.

EXAMPLES II

Put into the form $r(\cos \theta + i \sin \theta)$ the quantities

1. $1 + i$.

2. $-1 - i$.

3. $-\sqrt{3} + i$ ·

4. $3 + 4i$.

5. $1 + \sqrt{2} + i$.
Simplify

6. $2 - \sqrt{3} + i$.

7. $\dfrac{(\cos \theta - i \sin \theta)^{10}}{(\cos a + i \sin a)^{12}}$.

8. $\dfrac{(\cos a + i \sin a)(\cos \beta + i \sin \beta)}{(\cos \gamma + i \sin \gamma)(\cos \delta + i \sin \delta)}$.

9. $\dfrac{(\cos 2\theta - i \sin 2\theta)^7 (\cos 3\theta + i \sin 3\theta)}{(\cos 4\theta + i \sin 4\theta)^{12} (\cos 5\theta - i \sin 5\theta)^{-6}}$.

10. $\dfrac{\left(\cos \dfrac{\pi}{6} - i \sin \dfrac{\pi}{6}\right)^{11/2}}{\left(\cos \dfrac{\pi}{6} + i \sin \dfrac{\pi}{6}\right)^{1/2}}$.

11. $\dfrac{(\cos \alpha + i \sin \alpha)^1}{(\sin \beta + i \cos \beta)^5}$.

12. $\{(\cos\theta - \cos\phi) + i(\sin\theta - \sin\phi)\}^n + \{\cos\theta - \cos\phi - i(\sin\theta - \sin\phi)\}^n.$

13. Prove that

$$(\sin x + i\cos x)^n = \cos n\left(\frac{\pi}{2} - x\right) + i\sin n\left(\frac{\pi}{2} - x\right),$$

and that $\left(\dfrac{1+\sin\phi + i\cos\phi}{1+\sin\phi - i\cos\phi}\right)^n = \cos\left(\dfrac{n\pi}{2} - n\phi\right) + i\sin\left(\dfrac{n\pi}{2} - n\phi\right).$

If x, y, z and u stand respectively for

$\cos\alpha + i\sin\alpha,\ \cos\beta + i\sin\beta,\ \cos\gamma + i\sin\gamma,$ and $\cos\delta + i\sin\delta,$ prove that

14. $(x+y)(z+u) = 4\cos\dfrac{\alpha-\beta}{2}\cos\dfrac{\gamma-\delta}{2}\left[\cos\dfrac{\alpha+\beta+\gamma+\delta}{2} + i\sin\dfrac{\alpha+\beta+\gamma+\delta}{2}\right].$

15. $\dfrac{1}{(x-y)(z-u)} = -\dfrac{1}{4}\mathrm{cosec}\dfrac{\alpha-\beta}{2}\,\mathrm{cosec}\dfrac{\gamma-\delta}{2}\left[\cos\dfrac{\alpha+\beta+\gamma+\delta}{2} - i\sin\dfrac{\alpha+\beta+\gamma+\delta}{2}\right].$

16. $xy + zu = 2\cos\dfrac{\alpha+\beta-\gamma-\delta}{2}\left[\cos\dfrac{\alpha+\beta+\gamma+\delta}{2} + i\sin\dfrac{\alpha+\beta+\gamma+\delta}{2}\right].$

17. From the identity

$$(a^2 - b^2)(c^2 - d^2) = (c^2 - b^2)(a^2 - d^2) + (a^2 - c^2)(b^2 - d^2)$$

prove, by putting $a = \cos\alpha + i\sin\alpha$ and similar expressions for the other letters, the identity

$\sin(\alpha - \beta)\sin(\gamma - \delta) = \sin(\alpha - \delta)\sin(\gamma - \beta) + \sin(\alpha - \gamma)\sin(\beta - \delta).$

18. From the identity

$$\frac{(x-b)(x-c)}{(a-b)(a-c)} + \frac{(x-c)(x-a)}{(b-c)(b-a)} + \frac{(x-a)(x-b)}{(c-a)(c-b)} = 1$$

deduce, by assuming $x = \cos 2\theta + i\sin 2\theta$ and corresponding quantities for a, b and c, that

$$\frac{\sin(\theta - \beta)\sin(\theta - \gamma)}{\sin(\alpha - \beta)\sin(\alpha - \gamma)}\sin 2(\theta - \alpha) + \text{two similar expressions} = 0.$$

Similarly, deduce identities from the identity

$$\frac{1}{(x-a)(x-b)} = \frac{1}{(a-b)(x-a)} - \frac{1}{(a-b)(x-b)}.$$

19. Prove that

$$(a+bi)^{\frac{m}{n}} + (a-bi)^{\frac{m}{n}} = 2(a^2 + b^2)^{\frac{m}{2n}}\cos\left(\frac{m}{n}\tan^{-1}\frac{b}{a}\right).$$

20. If $2\cos\theta = x + \dfrac{1}{x},$

prove that $2\cos r\theta = x^r + \dfrac{1}{x^r}.$

21. If $2\cos\theta = x + \dfrac{1}{x},\ 2\cos\theta = y + \dfrac{1}{y},\ ......$

prove that $2\cos(\theta + \phi + ...) = xyz... + \dfrac{1}{xyz...}.$

22. If $x_r = \cos\dfrac{\pi}{2^r} + \sqrt{-1}\sin\dfrac{\pi}{2^r}$,

prove that $x_1, x_2, x_3, \ldots$ ad inf. $= \cos\pi$.

23. Using De Moivre's Theorem solve the equation,

$$x^4 - x^3 + x^2 - x + 1 = 0.$$

▮ **23.** In Art. 21 we have only shown that

$$\cos\frac{\theta}{q} + \sqrt{-1}\sin\frac{\theta}{q}$$

is one of the values of

$$\left(\cos\theta + \sqrt{-1}\sin\theta\right)^{1/q}.$$

The other values may be easily obtained. For

$$\left(\cos\theta + \sqrt{-1}\sin\theta\right)^{1/q} = \Big[\cos(2n\pi + \theta) + \sqrt{-1}$$

$\sin(2n\pi + \theta)\big]^{1/q}$, where n is any integer, and one of the values of the latter quantity is

$$\cos\frac{2n\pi + \theta}{q} + \sqrt{-1}\sin\frac{2n\pi + \theta}{q}.$$

By giving n the successive values 0, 1, 2, 3, ... $(q-1)$ we see that each of the quantities

$$\cos\frac{\theta}{d} + \sqrt{-1}\sin\frac{\theta}{q},$$

$$\cos\frac{2\pi + \theta}{q} + \sqrt{-1}\sin\frac{2\pi + \theta}{q},$$

$$\cos\frac{4\pi + \theta}{q} + \sqrt{-1}\sin\frac{4\pi + \theta}{q},$$

$$\cos\frac{6\pi + \theta}{q} + \sqrt{-1}\sin\frac{6\pi + \theta}{q} \tag{1}$$

$$\cdot\;\cdot\;\cdot\;\cdot\;\cdot\;\cdot\;\cdot\;\cdot\;\cdot\;\cdot\;\cdot\;\cdot\;\cdot$$

is equal to one of the values of

$$\left(\cos\theta + \sqrt{-1}\sin\theta\right)^{1/q}.$$

The highest value that we need assign to n is $q-1$; for the values q, $q+1$, $q+2$, ... will be found to give the same result as the values 0, 1, 2, ...

Also no two of the quantities (1) will be the same. For all the angles involved therein differ from one another by less than 2π and no two angles, differing by less than 2π, have their cosines the same and also their sines the same.

To sum up; By giving to n the successive values 0, 1, 2, ... $q-1$ in the expression

$$\cos\frac{2n\pi+\theta}{q}+\sqrt{-1}\sin\frac{2n\pi+\theta}{q}$$

we obtain q, and only q, different values for

$$\left(\cos\theta+\sqrt{-1}\sin\theta\right)^{1/q}.$$

■ **24.** By the use of the last article we can now obtain trigonometrical expressions for any root of a quantity which is of the form $x + yi$.

For we proved in Art. 20 that

$$x+yi=\rho[\cos(2n\pi+\theta)+\sqrt{-1}\sin(2n\pi+\theta)],$$

where
$$\rho=+\sqrt{x^2+y^2},$$

and θ is such that

$$\cos\theta=\frac{x}{\rho}\text{ and }\sin\theta=\frac{y}{\rho}.$$

Hence

$$(x+yi)^{1/q}=\rho^{1/q}\left[\cos\frac{2n\pi+\theta}{q}+\sqrt{-1}\sin\frac{2n\pi+\theta}{q}\right].$$

By giving n in succession the values 0, 1, 2, ... $q - 1$, we obtain the q required roots.

■ **25. EXAMPLE 1.** *Find the values of*

$$\left(\cos\frac{\pi}{3}+\sqrt{-1}\sin\frac{\pi}{3}\right)^{1/4}.$$

We have

$$\left(\cos\frac{\pi}{3}+\sqrt{-1}\sin\frac{\pi}{3}\right)^{1/4}=\left[\cos\left(2n\pi+\frac{\pi}{3}\right)+\sqrt{-1}\sin\left(2n\pi+\frac{\pi}{3}\right)\right]^{1/4},$$

where n is any integer,

$$=\cos\left(\frac{2n\pi}{4}+\frac{\pi}{12}\right)+\sqrt{-1}\sin\left(\frac{2n\pi}{4}+\frac{\pi}{12}\right).$$

Giving n in succession the values 0, 1, 2, and 3 we have as our answers the quantities

$$\cos\frac{\pi}{12}+\sqrt{-1}\sin\frac{\pi}{12},\ \cos\frac{7\pi}{12}+\sqrt{-1}\sin\frac{7\pi}{12},$$

$$\cos\frac{13\pi}{12}+\sqrt{-1}\sin\frac{13\pi}{12},\text{ and }\cos\frac{19\pi}{12}+\sqrt{-1}\sin\frac{19\pi}{12}.$$

The student will note that the value $n = 4$ will not give us an additional value. For it gives

$$\cos\left(2\pi+\frac{\pi}{12}\right)+\sqrt{-1}\sin\left(2\pi+\frac{\pi}{12}\right),$$

which is the same as $\cos\dfrac{\pi}{12}+\sqrt{-1}\sin\dfrac{\pi}{12},$

and this is the first of the quantities already found. Similarly the values $n = 5$, $n = 6$ and $n = 7$ would only give respectively the remaining three quantities, and so on.

■ **EXAMPLE 2.** *Find all the values of* $(-1)^{1/3}$.

Since
$$\cos \pi = -1, \text{ and } \sin \pi = 0,$$

we have
$$(-1)^{1/3} = (\cos \pi + \sqrt{-1} \sin \pi)^{1/3}$$

$$= \left[\cos(2n\pi + \pi) + \sqrt{-1} \sin(2n\pi + \pi)\right]^{1/3}$$

$$= \cos \frac{2n\pi + \pi}{3} + \sqrt{-1} \sin \frac{2n\pi + \pi}{3}.$$

Giving n the values 0, 1, and 2, the required values are

$$\cos \frac{\pi}{3} + \sqrt{-1} \sin \frac{\pi}{3}, \cos \pi + \sqrt{-1} \sin \pi, \text{ and } \cos \frac{5\pi}{3} + \sqrt{-1} \sin \frac{5\pi}{3}.$$

i.e.
$$\frac{1 + \sqrt{-3}}{2}, -1 \text{ and } \frac{1 - \sqrt{-3}}{2}.$$

■ **EXAMPLE 3.** *Solve the equation* $x^3 - x^5 + x^4 - 1 = 0$.

The equation is
$$(x^5 + 1)(x^4 - 1) = 0.$$

Taking the first factor, we have

$$x^5 = -1 = \cos(2r + 1)\pi + \sqrt{-1} \sin(2r + 1)\pi,$$

so that
$$x = \cos \frac{2r + 1}{5}\pi + \sqrt{-1} \sin \frac{2r + 1}{5}\pi.$$

Giving r the values 0, 1, 2, 3, 4 successively we have as solutions

$$\cos 36° + \sqrt{-1} \sin 36°, \cos 108° + \sqrt{-1} \sin 108°, \cos 180° + \sqrt{-1} \sin 180°,$$

$$\cos 252° + \sqrt{-1} \sin 252°.$$

and
$$\cos 324° + \sqrt{-1} \sin 324°.$$

Taking the second factor, we have

$$x^4 = 1 = \cos 2n\pi + \sqrt{-1} \sin 2n\pi,$$

so that
$$x = \cos \frac{n\pi}{2} + \sqrt{-1} \sin \frac{n\pi}{2}.$$

Giving n the values 0, 1, 2, and 3 we have solutions 1, $\sqrt{-1}$, -1, and $-\sqrt{-1}$. Hence all the roots are known.

EXAMPLES III

Find all the values of

1. $1^{1/3}$.

2. $(-1)^{1/5}$.

3. $(-i)^{1/6}$.

4. $(-1)^{1/10}$.

5. $(1 + \sqrt{-1})^{1/6}$.

6. $(1 + \sqrt{-3})^{11/3}$.

7. $\left(1-\sqrt{-3}\right)^{1/4}$.

8. $\left(\sqrt{3}+\sqrt{-1}\right)^{1/3}$.

9. $\left(\sqrt{3}-\sqrt{-1}\right)^{2/5}$.

10. $16^{1/4}$.

11. $32^{1/5}$.

12. $\left(1+\sqrt{-3}\right)^{10}+\left(1-\sqrt{-3}\right)^{10}$.

13. Simplify
$$\left(\cos\frac{2\pi}{3}+i\sin\frac{2\pi}{3}\right)^{1/4}$$
and express the results in a form free from trigonometrical expressions.

14. Find the continued product of the four values of
$$\left(\cos\frac{\pi}{3}+i\sin\frac{\pi}{3}\right)^{3/4}.$$

15. Prove that the roots of the equation $x^{10} + 11x^5 - 1 = 0$ are
$$\frac{\pm\sqrt{5}-1}{2}\left[\cos\frac{2r\pi}{5}+i\sin\frac{2r\pi}{5}\right].$$

16. Solve the equation $x^{12} - 1 = 0$ and find which of its roots satisfy the equation
$$x^4 + x^2 + 1 = 0.$$

Solve the equations

17. $x^7 + 1 = 0$.

18. $x^7 + x^4 + x^3 + 1 = 0$.

19. Prove that $\sqrt[n]{a+bi}+\sqrt[n]{a-bi}$

has n real values and find those of
$$\sqrt[3]{1+\sqrt{-3}}+\sqrt[3]{1-\sqrt{-3}}\ .$$

20. Prove that the nth roots of unity form a series in G.P.

21. Find the seven 7th roots of unity and prove that the sum of their nth powers always vanishes unless n be a multiple of 7, n being an integer, and that then the sum is 7.

■ **26. Binomial Theorem for Complex Quantities.**

It is known that for any real values of n and z, provided that z be less than unity, we have

$$(1+z)^n = 1+nz+\frac{n(n-1)}{1\cdot2}z^2+\frac{n(n-1)(n-2)}{1\cdot2\cdot3}z^3+... \qquad ...(1)$$

When z is complex $\left(=x+y\sqrt{-1}\right)$ and n is a positive integer, the ordinary proof applies and the theorem (1) is still true.

When z is complex, and n is a fraction or negative, it can be shown that

$$1+nz+\frac{n(n-1)}{1\cdot2}z^2+... \qquad ...(2)$$

is **one** of the values of $(1 + z)^n$, provided that the modulus of z, i.e. $\sqrt{x^2+y^2}$, is less than unity. When this modulus is equal to unity, the theorem is only true (1) when n is positive, and (2) when n is a negative proper fraction and z is not equal to -1.

The proof is difficult and beyond the range of the present book. We shall therefore assume the result. The student may hereafter refer to Hobson's *Trigonometry*, Arts. 211 and 212, or Chrystal's *Algebra*, Vol. II page 262.

CHAPTER 3

Expansions of sin $n\theta$ and cos $n\theta$. Series for sin θ and cos θ in Powers of θ

■ **27.** By the use of De Moivre's Theorem we can obtain the expansion of cos $n\theta$ and sin $n\theta$ in terms of the trigonometrical functions of θ.

For we have

$$\cos n\theta + i \sin n\theta = (\cos \theta + i \sin \theta)^n$$

Since n is a positive integer, the Binomial Theorem holds for $(\cos \theta + i \sin \theta)^n$.

Hence, by expanding, we have

$$\cos n\theta + i \sin n\theta = \cos^n \theta + n \cos^{n-1} \theta \cdot i \sin \theta$$

$$+ \frac{n(n-1)}{1 \cdot 2} \cos^{n-2} \theta \cdot i^2 \sin^2 \theta + \frac{n(n-1)(n-2)}{1 \cdot 2 \cdot 3} \cos^{n-3} \theta \cdot i^3 \sin^3 \theta ...$$

Hence, since

$$i^2 = -1, \ i^3 = -i, \ i^4 = 1, \ i^5 = i, \ ...$$

we have

$$\cos n\theta + i \sin n\theta = \cos^n \theta - \frac{n(n-1)}{1 \cdot 2} \cos^{n-2} \theta \sin^2 \theta$$

$$+ \frac{n(n-1)(n-2)(n-3)}{1 \cdot 2 \cdot 3 \cdot 4} \cos^{n-4} \theta \sin^4 \theta + ...$$

$$+ i \left[n \cos^{n-1} \theta \sin \theta - \frac{n(n-1)(n-2)}{1 \cdot 2 \cdot 3} \cos^{n-3} \theta \sin^3 \theta + ... \right].$$

By equating real and imaginary parts, we have

$$\cos n\theta = \cos^n \theta - \frac{n(n-1)}{1 \cdot 2} \cos^{n-2} \theta \sin^2 \theta + ... \qquad ...(1)$$

and

$$\sin n\theta = n \cos^{n-1} \theta \sin \theta - \frac{n(n-1)(n-2)}{1 \cdot 2 \cdot 3} \cos^{n-3} \theta \sin^3 0$$

$$+ \frac{n(n-1)(n-2)(n-3)(n-4)}{1 \cdot 2 \cdot 3 \cdot 4 \cdot 5} \cos^{n-5} \theta \sin^5 \theta - ... \qquad ...(2)$$

The terms in each of these series are alternately positive and negative. Also each series continues till one of the factors in the numerator is zero and then ceases.

■ **28.** From equations (1) and (2) of the last article we have, by division,

$$\tan n\theta = \frac{\sin n\theta}{\cos n\theta}$$

$$= \frac{n\cos^{n-1}\theta\sin\theta - \dfrac{n(n-1)(n-2)}{1\cdot 2\cdot 3}\cos^{n-2}\theta\sin^2\theta + \ldots}{\cos^n\theta - \dfrac{n(n-1)}{1\cdot 2}\cos^{n-2}\theta\sin^2\theta + \dfrac{n(n-1)(n-2)(n-3)}{1\cdot 2\cdot 3\cdot 4}\cos^{n-4}\theta\sin^4\theta\ldots}$$

Divide the numerator and denominator of the right-hand member of this equation by $\cos^n\theta$, and we have

$$\tan n\theta = \frac{n\tan\theta - \dfrac{n(n-1)(n-2)}{1\cdot 2\cdot 3}\tan^3\theta + \dfrac{n(n-1)(n-2)(n-3)(n-4)}{\underline{|5}}\tan^5\theta\ldots}{1 - \dfrac{n(n-1)}{1\cdot 2}\tan^2\theta + \dfrac{n(n-1)(n-2)(n-3)}{\underline{|4}}\tan^4\theta\ldots}$$

■ **29.** The values for $\cos n\theta$ and $\sin n\theta$ in Art. 27 may also be obtained, by Induction, without the use of imaginary quantities.

For assume (1) and (2) to be true for any value of n. Then, since

$$\cos (n + 1)\theta = \cos n\theta \cos \theta - \sin n\theta \sin \theta.$$

we obtain the value of $\cos (n + 1)\theta$, which after rearrangement, is found to be obtained from (1) by changing n into $(n + 1)$.

Similarly for $\sin (n + 1)\theta$.

Hence, if the formulae (1) and (2) are true for one value of n, they are true for the next greater value.

But it is easy to show that they are true for the values $n = 2$ and $n = 3$. Hence, by Induction, they can be proved to be true for all values of n.

■ **30.** From De Moivre's Theorem may be deduced expressions for the sine, cosine and tangent of the sum of any number of unequal angles in terms of the tangents of these angles.

For we have

$$\cos (\alpha + \beta + \gamma + \ldots) + i \sin (\alpha + \beta + \gamma + \ldots)$$
$$= (\cos \alpha + i \sin \alpha)(\cos \beta + i \sin \beta)(\cos \gamma + i \sin \gamma) \qquad \ldots(1)$$

Now　$\cos \alpha + i \sin \alpha = \cos \alpha [1 + i \tan \alpha],$

$\cos \beta + i \sin \beta = \cos \beta (1 + i \tan \beta),$

. . . .

Hence (1) may be written

$$\cos (\alpha + \beta + \gamma + \ldots) + i \sin (\alpha + \beta + \gamma + \ldots)$$
$$= \cos \alpha \cos \beta \cos \gamma \ldots (1 + i \tan \alpha)(1 + i \tan \beta)(1 + i \tan \gamma)\ldots$$
$$= \cos \alpha \cos \beta \cos \gamma \ldots[1 + i(\tan \alpha + \tan \beta + \tan \gamma + \ldots)$$
$$+ i^2(\tan \alpha \tan \beta + \tan \beta \tan \gamma + \ldots)$$
$$+ i^3 (\tan \alpha \tan \beta \tan \gamma + \tan \beta \tan \gamma \tan \delta \ldots)$$
$$+ \ldots] \ldots \ldots \ldots \ldots \ldots \ldots \ldots \qquad \ldots(2)$$

Using the notation of Art. 125 (Part I.), this equation may be written

$$\cos(\alpha + \beta + \gamma + ...) + i\sin(\alpha + \beta + \gamma + ...)$$
$$= \cos\alpha\cos\beta\cos\gamma ...[1 + is_1 - s_2 - is_3 + s_4 + is_5 - s_6 ...].$$

Hence equating real and imaginary parts, we have

$$\sin(\alpha + \beta + \gamma ...) = \cos\alpha\cos\beta\cos\gamma ...[s_1 - s_3 + s_5 - s_7 ...] \qquad ...(3),$$

and

$$\cos(\alpha + \beta + \gamma ...) = \cos\alpha\cos\beta\cos\gamma ...(1 - s_2 + s_4 - s_6 ...) \qquad ...(4)$$

Hence, by division,

$$\tan(\alpha + \beta + \gamma +...) = \frac{s_1 - s_3 + s_5 - s_7 ...}{1 - s_2 + s_4 - s_6 ...} \qquad ...(5)$$

The signs in the expressions on the right hand of (3) and (4) are alternately positive and negative.

The relation (5) was shown, by Induction, to be true in Part I., Art. 125.

▨ **31. EXAMPLE.** *Prove that the equation*

$$a^2\cos^2\theta + b^2\sin^2\theta + 2\,ga\cos\theta + 2\,fb\sin\theta + c = 0$$

has 4 roots, and that the sum of the values of θ which satisfy it is an even multiple of π radians.

Let $\quad t \equiv \tan\dfrac{\theta}{2}.$

Then since (Art. 109, Part I.), $\quad \sin\theta = \dfrac{2\tan\dfrac{\theta}{2}}{1 + \tan^2\dfrac{\theta}{2}} \quad$ and $\quad \cos\theta = \dfrac{1 - \tan^2\dfrac{\theta}{2}}{1 + \tan^2\dfrac{\theta}{2}},$

the equation above becomes

$$a^2\left(\frac{1-t^2}{1+t^2}\right)^2 + b^2\left(\frac{2t}{1+t^2}\right)^2 + 2ga\frac{1-t^2}{1+t^2} + 2fb\frac{2t}{1+t^2} + c = 0,$$

or, on reduction and simplification,

$$t^4(a^2 - 2ga + c) + 4fbt^3 + t^2(4b^2 - 2a^2 + 2c) + 4fbt + a^2 + 2ga + c = 0 \quad ...(1).$$

This is an equation having 4 roots.

Also, $\quad s_1 = $ sum of the roots $ = -\dfrac{4fb}{a^2 - 2ga + c},$

$\quad s_2 = $ sum taken two at a time $ = \dfrac{4b^2 - 2a^2 + 2c}{a^2 - 2ga + c},$

$\quad s_3 = $ sum taken three at a time $ = -\dfrac{4fb}{a^2 - 2ga + c},$

and $\quad s_4 = $ sum taken four at a time $ = \dfrac{a^2 + 2ga + c}{a^2 - 2ga + c}.$

Since, $s_1 = s_3$, it follows, by the last article, that

$$\tan\left(\frac{\theta_1 + \theta_2 + \theta_3 + \theta_4}{2}\right) = \frac{s_1 - s_3}{1 - s_2 + s_4} = 0 = \tan n\pi.$$

[The denominator $1 - s_2 + s_4$ does not vanish unless $a^2 = b^2$.]

$$\therefore\ \theta_1 + \theta_2 + \theta_3 + \theta_4 = 2 \cdot n\pi \text{ radians}$$

$$= \text{an even multiple of } \pi \text{ radians.}$$

[The student who is acquainted with Analytical Geometry will see that this is a solution of the problem. "If a circle and an ellipse intersect in four points, prove that the sum of the eccentric angles of the four points is equal to an even multiple of two right angles."]

EXAMPLES IV

Prove that

1. $\cos 4\theta = \cos^4\theta - 6 \cos^2\theta \sin^2\theta + \sin^4\theta.$

2. $\sin 6\theta = 6 \cos^5\theta \sin\theta - 20 \cos^2\theta \sin^3\theta + 6 \cos\theta \sin^5\theta.$

3. $\sin 7\theta = 7 \cos^6\theta \sin\theta - 35 \cos^4\theta \sin^3\theta + 21 \cos^2\theta \sin^5\theta - \sin^7\theta.$

4. $\cos 9\theta = \cos^9\theta - 36 \cos^7\theta \sin^2\theta + 126 \cos^5\theta \sin^4\theta - 84 \cos^3\theta \sin^6\theta + 9 \cos\theta \sin^8\theta.$

5. $\sin 8\theta = \cos^8\theta - 28 \cos^6\theta \sin^2\theta + 70 \cos^4\theta \sin^4\theta - 28 \cos^2\theta \sin^6\theta + \sin^8\theta.$

   Write down, in terms of $\tan\theta$, the values of

6. $\tan 5\theta.$　　　　　　　　7. $\tan 7\theta.$　　　　　　　　8. $\tan 9\theta.$

9. Prove that the last terms in the expressions for $\cos 11\theta$ and in 11θ are

$$-11 \cos\theta \sin^{10}\theta \text{ and } - \sin^{11}\theta.$$

10. Prove that the last terms in the expressions for $\sin 8\theta$ and $\sin 9\theta$ are $-8 \cos\theta \sin^7\theta$ and $\sin^9\theta$ respectively.

11. When n is odd, prove that the last terms in the expansions of $\sin n\theta$ and $\cos n\theta$ are respectively.

$$(-1)^{n-1/2} \sin^n\theta \text{ and } n(-1)^{n-1/2} \cos\theta\sin^{n-1}\theta.$$

12. When n is even, prove that the last terms in the expansion of $\sin n\theta$ and $\cos n\theta$ are respectively

$$n(-1)^{n-2/2} \cos\theta\sin^{n-1}\theta \text{ and } (-1)^{n/2} \sin^n\theta.$$

13. If α, β and γ be the roots of the equation

$$x^3 + px^2 + qx + p = 0,$$

prove that $\tan^{-1}\alpha + \tan^{-1}\beta + \tan^{-1}\gamma = n\pi$ radians except in one particular case.

14. Prove that the equation

$$\sin 3\theta = a \sin\theta + b \cos\theta + c$$

has six roots and that the sum of the six values of θ, which satisfy it, is equal to an odd multiple of π radians.

15. Prove that the equation

$$ah \sec\theta - bk \csc\theta = a^2 - b^2$$

has four roots, and that the sum of the four values of θ, which satisfy it, is equal to an odd multiple of π radians.

16. If θ_1, θ_2, θ_3 be three values of θ which satisfy the equation

$$\tan 2\theta = \lambda \tan (\theta + \alpha),$$

and such that no two of them differ by a multiple of π, show that $\theta_1 + \theta_2 + \theta_3 + \alpha$ is a multiple of π.

Expansions of the sine and cosine of an angle in series of ascending powers of the angle.

■ **32.** As in Art. 27 we have

$$\cos n\theta = \cos^n \theta - \frac{n(n-1)}{1 \cdot 2} \cos^{n-2} \theta \sin^2 \theta$$

$$+ \frac{n(n-1)(n-2)(n-3)}{1 \cdot 2 \cdot 3 \cdot 4} \cos^{n-4} \theta \sin^4 \theta - \dots$$

Put $n\theta = \alpha$ and we have

$$\cos \alpha = \cos^n \theta - \frac{\dfrac{\alpha}{\theta}\left(\dfrac{\alpha}{\theta} - 1\right)}{1 \cdot 2} \cos^{n-2} \theta \sin^2 \theta$$

$$+ \frac{\dfrac{\alpha}{\theta}\left(\dfrac{\alpha}{\theta} - 1\right)\left(\dfrac{\alpha}{\theta} - 2\right)\left(\dfrac{\alpha}{\theta} - 3\right)}{1 \cdot 2 \cdot 3 \cdot 4} \cos^{n-4} \theta \sin^4 \theta - \dots$$

$$= \cos^n \theta - \frac{\alpha(\alpha - \theta)}{1 \cdot 2} \cos^{n-2} \theta \left(\frac{\sin\theta}{\theta}\right)^2$$

$$+ \frac{\alpha(\alpha - \theta)(\alpha - 2\theta)(\alpha - 3\theta)}{1 \cdot 2 \cdot 3 \cdot 4} \cos^{n-4} \theta \left(\frac{\sin\theta}{\theta}\right)^4 - \qquad \dots(1)$$

In equation (1) make θ indefinitely small, α remaining constant and therefore n becoming indefinitely great.

Then $\dfrac{\sin\theta}{\theta}$ is, in the limit, equal to unity and so is every power of $\left(\dfrac{\sin\theta}{\theta}\right)$. (Art. 15.)

Also $\cos \theta$ is, in the limit, equal to unity and so also is every power of $\cos \theta$. (Art. 14.)

Hence (1) becomes

$$\cos \alpha = 1 - \frac{\alpha^2}{\lfloor 2} + \frac{\alpha^4}{\lfloor 4} - \frac{\alpha^6}{\lfloor 6} + \dots \text{ ad inf.}$$

[Cf. Art. 16.]

■ **33.** *To expand sin α in terms of α.*

As in Art. 27, we have

$$\sin n\theta = n\cos^{n-1} \theta \sin\theta - \frac{n(n-1)(n-2)}{1 \cdot 2 \cdot 3} \cos^{n-3} \theta \sin^3 \theta + \dots$$

As before put $n\theta = \alpha$, and we have

$$\sin a = \frac{\alpha}{\theta}\cos^{n-1}\theta\sin\theta - \frac{\frac{\alpha}{\theta}\left(\frac{\alpha}{\theta}-1\right)\left(\frac{\alpha}{\theta}-2\right)}{1\cdot 2\cdot 3}\cos^{n-3}\theta\sin^3\theta$$

$$+\frac{\frac{\alpha}{\theta}\left(\frac{\alpha}{\theta}-1\right)\left(\frac{\alpha}{\theta}-2\right)\left(\frac{\alpha}{\theta}-3\right)\left(\frac{\alpha}{\theta}-4\right)}{1\cdot 2\cdot 3\cdot 4\cdot 5}\cos^{n-5}\theta\sin^5\theta +...$$

$$= \alpha\cos^{n-1}\theta\cdot\left(\frac{\sin\theta}{\theta}\right) - \frac{\alpha(\alpha-\theta)(\alpha-2\theta)}{1\cdot 2\cdot 3}\cos^{n-3}\theta\left(\frac{\sin\theta}{\theta}\right)^3 +...$$

As in the last article make θ indefinitely small, keeping a finite, and we have

$$\sin\alpha = \alpha - \frac{\alpha^3}{\underline{|3}} + \frac{\alpha^5}{\underline{|5}} - \frac{\alpha^7}{\underline{|7}} +... \text{ ad inf.}$$

[Cf. Art. 16.]

■ **34.** There is no series, proceeding according to a simple law, for the expansion of $\tan\theta$ in terms of θ, similar to those of Arts. 32 and 33.

We shall find the series for $\tan\theta$ as far as the term involving θ^5.

For
$$\tan\theta = \frac{\sin\theta}{\cos\theta} = \frac{\theta - \dfrac{\theta^3}{\underline{|3}} + \dfrac{\theta^5}{\underline{|5}} -...}{1 - \dfrac{\theta^2}{\underline{|2}} + \dfrac{\theta^4}{\underline{|4}} -...}$$

$$= \left(\theta - \frac{\theta^3}{6} + \frac{\theta^5}{120} -...\right)\left[1 - \left(\frac{\theta^2}{2} - \frac{\theta^4}{24} +...\right)\right]^{-1}$$

$$= \left(\theta - \frac{\theta^3}{6} + \frac{\theta^5}{120} -...\right)\left[1 + \left(\frac{\theta^2}{2} - \frac{\theta^4}{24} +...\right) + \left(\frac{\theta^2}{2} - \frac{\theta^4}{24}...\right)^2...\right],$$

by the Binomial Theorem,

$$= \left(\theta - \frac{\theta^3}{6} + \frac{\theta^5}{120} -...\right)\left[1 + \frac{\theta^2}{2} - \frac{\theta^4}{24}...+\frac{\theta^4}{4}...\right],$$

neglecting θ^6 and higher powers of θ,

$$= \left(\theta - \frac{\theta^3}{6} + \frac{\theta^5}{120} -...\right)\left(1 + \frac{\theta^2}{2} + \frac{5}{24}\theta^4...\right)$$

$$= \theta + \frac{\theta^3}{3} + \frac{2}{15}\theta^5,$$

on reduction and neglecting powers of θ above θ^5.

A similar method would give the series for $\tan\theta$ to as many terms as we please. The method however soon becomes very cumbrous and troublesome.

▓ **35.** In Art. 32 and 33 we tacitly assumed that α was equal to the number of radians in the angle considered. For, unless this be the case, the limit of $\dfrac{\sin\theta}{\theta}$ is not unity when θ is made indefinitely small.

When the angle is expressed in degrees we proceed as follows.

Let $\quad \alpha° = x$ radians, so that

$$\frac{\alpha}{180} = \frac{x}{\pi},$$

and hence $\qquad x = \dfrac{\pi}{180}\alpha.$

Then $\qquad \cos\alpha° = \cos x°$

$$= 1 - \frac{x^2}{\underline{2}} + \frac{x^4}{\underline{4}} - \frac{x^6}{\underline{6}} + \ldots$$

$$= 1 - \frac{1}{\underline{2}}\frac{\pi^2\alpha^2}{180^2} + \frac{1}{\underline{4}}\frac{\pi^4\alpha^4}{180^4} - \frac{1}{\underline{6}}\frac{\pi^6\alpha^6}{180^6} + \ldots \text{ ad inf.}$$

So also

$$\sin\alpha° = \sin x^c = x - \frac{x^3}{\underline{3}} + \frac{x^5}{\underline{5}}\ldots$$

$$= \frac{\pi\alpha}{180} - \frac{1}{\underline{3}}\left(\frac{\pi\alpha}{180}\right)^3 + \frac{1}{\underline{5}}\left(\frac{\pi\alpha}{180}\right)^5 - \ldots \text{ad inf.}$$

▓ **36. Sines and cosines of small angles.** The series of Arts. 32 and 33 may be used to find the sines and cosines of small angles.

For example, let us find the values of $\sin 10^n$ and $\cos 10^n$.

Since $\qquad 10^n = \left(\dfrac{1}{6\times 60} \times \dfrac{\pi}{180}\right)$ radians

$$= \left(\frac{\pi}{64800}\right)^c,$$

we have

$$\sin 10^n = \frac{\pi}{64800} - \frac{1}{\underline{3}}\left(\frac{\pi}{64800}\right)^3 + \frac{1}{\underline{5}}\left(\frac{\pi}{64800}\right)^5 - \ldots$$

and $\qquad \cos 10^n = 1 - \dfrac{1}{\underline{2}}\left(\dfrac{\pi}{64800}\right)^2 + \dfrac{1}{\underline{4}}\left(\dfrac{\pi}{64800}\right)^4 - \ldots$

Now, $\qquad \dfrac{\pi}{64800} = 0.000048481368\ldots,$

$$\left(\frac{\pi}{64800}\right)^4 = 0.0000000023504...,$$

and

$$\left(\frac{\pi}{64800}\right)^3 = 0.00000000000113928.....$$

Hence, to twelve places of decimals, we have

$$\sin 10^n = 0.000048481368,$$

and

$$\cos 10^n = 1 - \frac{0.0000000023504}{2}$$

$$= 1 - 0.000000001175$$

$$= 0.999999998825.$$

■ **37. Approximate value of the root of an equation.** The series of Art. 33 may also be used to find an approximate value of the root of an equation. The method will be best shown by examples.

■ **EXAMPLE 1.** *If* $\dfrac{\sin\theta}{\theta} = \dfrac{1349}{1350}$, *prove that the angle* θ *is very nearly equal to* $\dfrac{1}{15}$ *th radians.*

We know that, the smaller θ is, the more nearly is $\dfrac{\sin\theta}{\theta}$ equal to unity. Conversely in our case we see that θ is small.

In the series for sin θ (Art. 33) let us omit the powers of θ above the third, and we have

$$\frac{\theta - \dfrac{\theta^3}{\underline{3}}}{\theta} = \frac{1349}{1350} = 1 - \frac{1}{1350}.$$

$$\therefore \qquad \theta^2 = \frac{6}{1350} = \frac{1}{225}.$$

Hence $\theta = \dfrac{1}{15}$, so that the angle is $\dfrac{1}{15}$ of a radian nearly.

If we desire a nearer approximation, we take the series for sin θ and omit powers above the 5th. We then have

$$\frac{\theta - \dfrac{\theta^3}{\underline{3}} + \dfrac{\theta^5}{\underline{5}}}{\theta} = 1 - \frac{1}{1350}.$$

This gives

$$\theta^4 - 20\,\theta^2 = -\frac{120}{1350} = -\frac{20}{225}.$$

Hence, by solving,

$$\theta^3 = 10 \pm \frac{\sqrt{22480}}{15} = \frac{150 - 149.933312...}{15} = \frac{0.66688}{15}$$

$$= \frac{1.00032}{15^2}.$$

$$\therefore \quad \theta = \frac{1.00016}{15} \text{ radian.}$$

This differs from the first approximation by about $\frac{1}{6000}$ th part.

■ **EXAMPLE 2.** *Solve approximately the equation*

$$\cos\left(\frac{\pi}{3} + \theta\right) = 0.49.$$

Since 0.49 is very nearly equal to $\frac{1}{2}$, which is the value of $\cos\frac{\pi}{3}$, it follows that θ must be small.

The equation may be written

$$\frac{1}{2}\cos\theta - \frac{\sqrt{3}}{2}\sin\theta = 0.49 = \frac{1}{2} - \frac{1}{100} \qquad \qquad ...(1)$$

For a first approximation omit squares and higher powers of θ. By Art. 33. this equation then becomes.

$$\frac{1}{2} \cdot 1 - \frac{\sqrt{3}}{2} \cdot \theta = \frac{1}{2} - \frac{1}{100},$$

so that

$$\theta = \frac{2}{\sqrt{3}} \frac{1}{100} = \frac{2\sqrt{3}}{300} = \frac{3.4641...}{300} = 0.011547... \text{ radian.}$$

For a still nearer approximation, omit cubes and higher powers of θ. The equation (1), then becomes

$$\frac{1}{2}\left(1 - \frac{\theta^2}{2}\right) - \frac{\sqrt{3}}{2}\theta = \frac{1}{2} - \frac{1}{100},$$

i.e.

$$\theta^2 + 2\sqrt{3}\theta = \frac{4}{100}.$$

$$\therefore \quad \theta = -\sqrt{3} + \frac{\sqrt{304}}{10} = 0.0115086... \text{ radian.}$$

The first approximation is therefore correct to 4 places of decimals.

The angle θ is therefore very nearly equal to 0.0115 radian, *i.e.* to about 40′.

The accurate answer is found, from the tables, to be 0.0115075... radian.

■ **38. Evaluation of quantities apparently indeterminate.** We often have to obtain the value of quantities which are apparently indeterminate.

Suppose we required the value of the expression

$$\frac{3\sin\theta - \sin 3\theta}{\theta(\cos\theta - \cos 3\theta)},$$

when θ is zero.

If we substitute the value 0 for θ, we have

$$\frac{0-0}{0\times 0},$$

which is apparently indeterminate.

The expression however, for all values of θ,

$$= \frac{3\sin\theta - (3\sin\theta - 4\sin^3\theta)}{\theta\{\cos\theta - (4\cos^3\theta - 3\cos\theta)\}} = \frac{4\sin^3\theta}{\theta\{4\cos\theta - 4\cos^3\theta\}}$$

$$= \frac{\sin^3\theta}{\theta\cos\theta\sin^2\theta} = \frac{\sin\theta}{\theta\cos\theta} = \frac{1}{\cos\theta}\times\frac{\sin\theta}{\theta}.$$

Now, the smaller θ is, the more nearly do both

$$\frac{1}{\cos\theta} \quad \text{and} \quad \frac{\sin\theta}{\theta}$$

approach to unity. Hence, when θ approaches the limit zero, the given expression approaches the limit 1×1, *i.e.* 1.

Such an expression as the one we have discussed is said to be indeterminate. We should more properly say that the expression is "at first sight" indeterminate.

■ **39.** In many cases the real value is very easily found by using the series for $\sin\theta$ and $\cos\theta$. The method is shown in the following examples, of the first of which the example in the preceding article is a particular case.

❙ **EXAMPLE 1.** *Find the value of*

$$\frac{n\sin\theta - \sin n\theta}{\theta(\cos\theta - \cos n\theta)} \quad \text{when } \theta \text{ is zero.}$$

The expression

$$= \frac{n\left(\theta - \dfrac{\theta^3}{\lfloor 3} + \dfrac{\theta^5}{\lfloor 5} - \cdots\right) - \left(n\theta - \dfrac{n^3\theta^3}{\lfloor 3} + \dfrac{n^5\theta^5}{\lfloor 5} - \cdots\right)}{\theta\left[\left(1 - \dfrac{\theta^2}{\lfloor 2} + \dfrac{\theta^4}{\lfloor 4}\cdots\right) - \left(1 - \dfrac{n^2\theta^2}{\lfloor 2} + \dfrac{n^4\theta^4}{\lfloor 4}\cdots\right)\right]}$$

$$= \frac{\dfrac{n^3-n}{\lfloor 3}\theta^2 - \dfrac{n^5-n}{\lfloor 5}\theta^5 + \text{higher powers of }\theta}{\theta\left[\dfrac{n^2-1}{\lfloor 2}\theta^2 - \dfrac{n^4-1}{\lfloor 4}\theta^4 + \text{higher powers of }\theta\right]}$$

$$= \dfrac{\dfrac{n^3 - n}{\lfloor\underline{3}} - \dfrac{n^5 - n}{\lfloor\underline{5}}\,\theta^2 + \text{higher powers}}{\dfrac{n^2 - 1}{\lfloor\underline{2}} - \dfrac{n^4 - 1}{\lfloor\underline{4}}\,\theta^2 + \text{higher powers}}$$

When θ is zero, this expression

$$= \dfrac{n^3 - n}{\lfloor\underline{3}} \div \dfrac{n^2 - 1}{\lfloor\underline{2}} = \dfrac{n}{3}.$$

▌ **EXAMPLE 2.** *Find the value, when x is zero, of the expression*

$$\frac{\cos x - \log_e(1 + x) + \sin x - 1}{e^x - (1 + x)}$$

Since

$$\log_e(1 + x) = x - \frac{1}{2}x^2 + \frac{1}{3}x^3 - \frac{1}{4}x^4 \ldots,$$

and

$$e^x = 1 + x + \frac{x^2}{\lfloor\underline{2}} + \frac{x^3}{\lfloor\underline{3}} + \frac{x^4}{\lfloor\underline{4}} \ldots \text{(Arts. 5 and 8)},$$

this expression

$$= \frac{\left(1 - \dfrac{x^2}{\lfloor\underline{2}} + \dfrac{x^4}{\lfloor\underline{4}} \ldots\right) - \left(x - \dfrac{1}{2}x^2 + \dfrac{1}{3}x^3 \ldots\right) + \left(x - \dfrac{x^3}{\lfloor\underline{3}} + \dfrac{x^5}{\lfloor\underline{5}} \ldots\right) - 1}{\left(1 + x + \dfrac{x^2}{\lfloor\underline{2}} + \dfrac{x^3}{\lfloor\underline{3}} + \ldots\right) - (1 + x)}$$

$$= \frac{-\dfrac{x^2}{\lfloor\underline{2}} + \text{higher powers of } x}{\dfrac{x^2}{\lfloor\underline{2}} + \text{higher powers of } x} = \frac{-\dfrac{x}{2} + \text{powers of } x}{\dfrac{1}{\lfloor\underline{2}} + \text{powers of } x}$$

when x is zero, this latter expression

$$= \frac{0}{1} = 0.$$

▌ **EXAMPLE 3.** *Find the value, when x is zero, of*

$$\left(\frac{\tan x}{x}\right)^{\frac{1}{x}},$$

when x is zero, this expression is of the form $\left(\dfrac{0}{0}\right)^{\infty}$

But it also

$$= \left(\frac{x + \dfrac{x^3}{3} + \ldots}{x}\right)^{1/x} \quad \text{(Art. 34).}$$

Now, by Art. 2, Cor., the value of

$$\left(1+\frac{x^2}{3}\right)^{\frac{3}{x^2}}$$

is e, when x is zero.

Hence, the expression $= e^{x/3} = e^0 = 1$

The value of the expression may be also found by finding the value of its logarithm.

EXAMPLES V

1. If
$$\frac{\sin\theta}{\theta} = \frac{1013}{1014},$$
   prove that θ is the number of radians is $4°\ 24'$ nearly.

2. If
$$\frac{\sin\theta}{\theta} = \frac{863}{864},$$
   prove that θ is equal to $4°\ 47'$ nearly.

3. If
$$\frac{\sin\theta}{\theta} = \frac{5045}{5046},$$
   prove that the angle θ is $1°\ 58'$ nearly.

4. If
$$\frac{\sin\theta}{\theta} = \frac{2165}{2166},$$
   prove that θ is equal to $3°\ 1'$ nearly.

5. If
$$\frac{\sin\theta}{\theta} = \frac{19493}{19494},$$
   prove that θ is equal to $1°$ nearly.

6. If
$$\tan\theta = \frac{1}{15},$$
   find an approximate value for θ.

   Find the value, when x is zero, of the expressions

7. $\dfrac{x-\sin x}{x^3}.$

8. $\dfrac{x^2}{1-\cos mx}.$

9. $\dfrac{\sin ax}{\sin bx}.$

10. $\dfrac{\tan x - \sin x}{\sin^3 x}.$

11. $\dfrac{\tan 2x - 2\sin x}{x^3}.$

12. $\dfrac{\text{ver}\sin ax_n}{\text{versin } bx}.$

13. $\dfrac{m\sin x - \sin mx}{m(\cos x - \cos mx)}.$

14. $\dfrac{a^2\sin ax - b^2\sin bx}{b^2\tan ax - a^2\tan bx}.$

15. $\dfrac{b^2\sin^2 ax - a^2\sin^2 bx}{b^2\tan^2 ax - a^2\tan^2 bx}.$

16. $\dfrac{x\log_e(1+x)}{1-\cos x}.$

17. $\dfrac{e^x - 1 + \log_e(1-x)}{\sin^3 x}.$

18. $\dfrac{x + 2\sin x - \sin 3x}{x + \tan x - \tan 2x}.$

19. $\dfrac{\sin x + \sin 6x - 7x}{x^4}.$

20. $\dfrac{\sin^2 nx - \sin^2 mx}{1 - \cos px}.$

21. $\dfrac{1}{x^4}\left[\dfrac{\sin x}{x} + \dfrac{e^x - e^{-x}}{2x} - 2\right].$

22. $\dfrac{\sin^2 \sqrt{mnx} - \sin mx \sin nx}{(1 - \cos mx)(1 - \cos nx)}.$

23. $\dfrac{3\sin x - \sin 3x}{x - \sin x}.$

24. $\dfrac{\left(\sin x - 2\sin \dfrac{x^2}{2}\right) + (1 - \cos x)}{\sin x \sin 2x - 8\cos x \sin^2 \dfrac{x}{2} - \dfrac{4}{3}\sin^4 x}.$

25. $\dfrac{a^x - b^x}{x}.$

26. $\left(\dfrac{\tan x}{x}\right)^{3/x^2}.$

27. $\left(\cos \dfrac{x}{m} + \sin \dfrac{3x}{m}\right)^{m/x}.$

Find the value, when x equals $\dfrac{\pi}{2}$, of

28. $\dfrac{(\cos x + \sin 2x + \cos 3x)^2}{(\sin x + 2\cos 2x - \sin 3x)^3}.$

29. $(\sin x)^{\tan x}.$

30. $\sec x - \tan x.$

Find the value, when n is infinite, of

31. $\left(\cos \dfrac{x}{n}\right)^n.$

32. $\left(\cos \dfrac{x}{n}\right)^{n^2}.$

33. $\left(\cos \dfrac{x}{n}\right)^{n^3}.$

34. If n be > 1 and $\theta = \dfrac{\pi}{2}$ nearly, prove that $(\sin \theta)^{1/n}$ is very nearly equal to

$\dfrac{(n-1) + (n+1)\sin \theta}{(n+1) + (n-1)\sin \theta}.$

35. In the limit, when $\beta = \alpha$, prove that

$\dfrac{\alpha \sin \beta - \beta \sin \alpha}{\alpha \cos \beta - \beta \cos \alpha} = \tan(\alpha - \tan^{-1}\alpha).$

36. Prove that

$4\tan^{-1}\dfrac{1}{5} - \dfrac{\pi}{4} = \tan^{-1}\dfrac{1}{239},$

and deduce that in a triangle ABC, in which C is a right angle and CA is five times CB, the angle A exceeds the eight part of a right angle by $6' 36''$, correct to the nearest second.

37. Find a and b so that the expression $a \sin x + b \sin 2x$ may be as close an approximation as possible to the number of radians in the angle x, when x is small.

38. If $y = x - e \sin x$, where e is very small, prove that

$$\tan \frac{y}{2} = \tan \frac{x}{2} \left(1 - e + e^2 \sin^2 \frac{x}{2} \right),$$

and that

$$\tan \frac{x}{2} = \tan \frac{y}{2} \left(1 + e + e^2 \cos^2 \frac{y}{2} \right),$$

where powers of e above the second are neglected.

39. If in the equation $\sin (\omega - \theta) = \sin \omega \cos \alpha$, θ be very small, prove that its approximate value is

$$2 \tan \omega \sin^2 \frac{\alpha}{2} \left(1 - \tan^2 \omega \sin^2 \frac{\alpha}{2} \right).$$

40. If ϕ be known by means of $\sin \phi$ to be an angle not $> 15°$, prove that its value differs from the fraction

$$\frac{28 \sin 2\phi + \sin 4\phi}{12 (3 + 2 \cos 2\phi)}$$

by less than the number of radians in $1'$.

▮ **40. EXAMPLE.** *Prove that the roots of the equation*

$$8x^3 - 4x^2 - 4x + 1 = 0 \qquad \qquad ...(1)$$

are $\qquad \qquad \cos \dfrac{\pi}{7}, \cos \dfrac{3\pi}{7}, \text{ and } \cos \dfrac{5\pi}{7},$

and hence that $\qquad \cos \dfrac{\pi}{7} + \cos \dfrac{3\pi}{7} + \cos \dfrac{5\pi}{7} = \dfrac{1}{2} \qquad \qquad ...(2)$

$$\cos \frac{\pi}{7} + \cos \frac{3\pi}{7} + \cos \frac{3\pi}{7} \cos \frac{5\pi}{7} + \cos \frac{5\pi}{7} \cos \frac{5\pi}{7} = -\frac{1}{2} \qquad ...(3)$$

and $\qquad \qquad \cos \dfrac{\pi}{7} \cos \dfrac{3\pi}{7} \cos \dfrac{5\pi}{7} = -\dfrac{1}{8} \qquad \qquad ...(4)$

First method. Let $y = \cos \theta + i \sin \theta$, where θ has either of the values

$$\frac{\pi}{7}, \frac{3\pi}{7}, \frac{5\pi}{7}, \pi, \frac{9\pi}{7}, \frac{11\pi}{7}, \text{ and } \frac{13\pi}{7}.$$

Then $\qquad \qquad y^7 = \cos 7\theta + i \sin 7\theta = -1,$

i.e. $\qquad \qquad (y + 1)(y^6 - y^5 + y^4 - y^3 + y^2 - y + 1) = 0.$

Now the root $y = -1$ corresponds to the value $\theta = \pi$.

The roots of the equation

$$y^6 - y^5 + y^4 - y^3 + y^2 - y + 1 = 0 \qquad \qquad ...(5)$$

are therefore $\cos \theta + i \sin \theta$, where θ has either of the values

$$\frac{\pi}{7}, \frac{3\pi}{7}, \frac{5\pi}{7}, \frac{9\pi}{7}, \frac{11\pi}{7}, \text{ or } \frac{13\pi}{7}.$$

Put
$$2x = y + \frac{1}{y} = \cos\theta + i\sin\theta + \frac{1}{\cos\theta + i\sin\theta}$$

$$= \cos\theta + i\sin\theta + \cos\theta - i\sin\theta = 2\cos\theta,$$

so that
$$y^2 + \frac{1}{y^2} = \left(y + \frac{1}{y}\right)^2 - 2 = 4x^2 - 2,$$

and
$$y^3 + \frac{1}{y^3} = \left(y + \frac{1}{y}\right)\left\{\left(y + \frac{1}{y}\right)^2 - 3\right\} = 8x^3 - 6x.$$

On dividing equation (5) by y^3 it becomes

$$y^3 + \frac{1}{y^3} - \left(y^2 + \frac{1}{y^2}\right) + \left(y + \frac{1}{y}\right) - 1 = 0,$$

i.e.
$$8x^3 - 4x^2 - 4x + 1 = 0 \qquad \qquad ...(6)$$

The roots of this equation are

$$\cos\frac{\pi}{7}, \cos\frac{3\pi}{7}, \cos\frac{5\pi}{7}, \cos\frac{9\pi}{7}, \cos\frac{11\pi}{7}, \text{ and } \cos\frac{13\pi}{7}.$$

Since
$$\cos\frac{13\pi}{7} = \cos\frac{\pi}{7}, \cos\frac{11\pi}{7} = \cos\frac{3\pi}{7},$$

and
$$\cos\frac{9\pi}{7} = \cos\frac{5\pi}{7},$$

the roots of (6) are therefore

$$\cos\frac{\pi}{7}, \cos\frac{3\pi}{7}, \text{and } \cos\frac{5\pi}{7}.$$

We then have

$$\cos\frac{\pi}{7} + \cos\frac{3\pi}{7} + \cos\frac{5\pi}{7} = \frac{4}{8} = \frac{1}{2},$$

$$\cos\frac{\pi}{7}\cos\frac{3\pi}{7} + \cos\frac{3\pi}{7} + \cos\frac{5\pi}{7} + \cos\frac{5\pi}{7}\cos\frac{\pi}{7} = \frac{-4}{8} = -\frac{1}{2},$$

and
$$\cos\frac{\pi}{7}\cos\frac{3\pi}{7}\cos\frac{5\pi}{7} = -\frac{1}{8}.$$

Second method. The equation
$$(\cos\theta + i\sin\theta)^7 = -1 \qquad \qquad ...(7)$$
i.e.
$$\cos 7\theta + i\sin 7\theta = -1$$
is clearly satisfied when θ has either of the values

$$\frac{\pi}{7}, \frac{3\pi}{7}, \frac{5\pi}{7}, \pi, \frac{9\pi}{7}, \frac{11\pi}{7}, \text{ and } \frac{13\pi}{7} \qquad \qquad ...(8)$$

Writing c for cos θ and s for sin θ, the equation (7), on being expanded by the Binomial Theorem, becomes

$$c^7 + 7ic^6s - 21c^5s^2 - 35ic^4s^3 + 35c^2s^4 + 21ic^2s^5 - 7cs^6 - is^7 = -1.$$

Equating the real parts on each side, we have

$$c^7 - 21c^5s^2 + 35c^3s^4 - 7cs^6 = -1.$$

Putting $s^2 = 1 - c^2$, we see that the cosine of each of the angles (8) satisfies the equation

$$64c^7 - 112c^5 + 56c^2 - 7c + 1 = 0 \qquad \qquad ...(9)$$

i.e. $\qquad\qquad (c + 1)\{8c^3 - 4c^2 - 4c + 1\}^2 = 0 \qquad \qquad ...(10)$

But

$$\cos \pi = -1, \cos\frac{13\pi}{7} = \cos\frac{\pi}{7}, \cos\frac{11\pi}{7} = \cos\frac{3\pi}{7} \text{ and } \cos\frac{9\pi}{7} = \cot\frac{5\pi}{7},$$

so that the roots of (10) are -1 and $\cos\dfrac{\pi}{7}$, $\cos\dfrac{3\pi}{7}$, and $\cos\dfrac{5\pi}{7}$, the latter three being

twice repeated.

Hence $\cos\dfrac{\pi}{7}$, $\cos\dfrac{3\pi}{7}$ and $\cos\dfrac{5\pi}{7}$ are the roots of the equation

$$8c^3 - 4c^2 - 4c + 1 = 0.$$

But this is equation (6).

The equation (9) may also be obtained by putting $\pi = 7$ in equation (2) of Art. 49, which is in the next chapter.

Third method. When only a small number of angles are introduced the equation (6) may be easily obtained without using imaginary quantities.

Let θ denote any of the angles (8).

Then $7\theta =$ an odd multiple of π.

∴ $\qquad\qquad\qquad \cos 4\theta = -\cos 3\theta,$

i.e. if $\qquad\qquad\qquad \cos \theta \equiv c$, we have

$$2\{2c^2 - 1\}^2 - 1 = -\{4c^3 - 3c\},$$

i.e. $\qquad\qquad 8c^4 - 8c^2 + 1 = 3c - 4c^3,$

i.e. $\qquad\qquad\quad 8c^4 + 4c^3 - 8c^2 - 3c + 1 = 0,$

i.e. $\qquad\qquad\quad (c + 1)(8c^3 - 4c^2 - 4c + 1) = 0.$

Hence as in the Second Method the roots of

$$8c^3 - 4c^2 - 4c + 1 = 0$$

are $\qquad\qquad \cos\dfrac{\pi}{7}, \cos\dfrac{3\pi}{7}, \text{ and } \cos\dfrac{5\pi}{7}$

■ **41.** From the preceding article we can obtain an equation giving

$$\sec^2\frac{\pi}{7}, \sec^2\frac{3\pi}{7}, \text{ and } \sec^2\frac{5\pi}{7}.$$

In equation (6) of that article put $\dfrac{1}{x^2} = y$, and therefore $x = \dfrac{1}{\sqrt{y}}$. It then follows that the quantities

$$\sec^2\frac{\pi}{7}, \; \sec^2\frac{3\pi}{7}, \; \text{and} \; \sec^2\frac{5\pi}{7}$$

are the roots of the equation

$$8\frac{1}{y\sqrt{y}} - \frac{4}{y} - \frac{4}{\sqrt{y}} + 1 = 0,$$

or, on rationalizing,

$$y^3 - 24y^2 + 80y - 64 = 0 \qquad \qquad \text{...(1)}$$

Again, putting $y = 1 + z$, then, since $\sec^2\theta = 1 + \tan^2\theta$, it follows that

$$\tan^2\frac{\pi}{7}, \; \tan^2\frac{3\pi}{7} \; \text{and} \; \tan^2\frac{5\pi}{7}$$

are the roots of the equation

$$(1 + z)^3 - 24(1 + z)^2 + 80(1 + z) - 64 = 0,$$

i.e. $\qquad\qquad z^3 - 21z^2 + 35z - 7 = 0 \qquad\qquad \text{...(2)}.$

The equation (2) may be easily obtained directly.

For, if θ stand for either of the angles

$$\frac{\pi}{7}, \frac{2\pi}{7}, \frac{3\pi}{7}, \frac{4\pi}{7}, \frac{5\pi}{7}, \frac{6\pi}{7} \; \text{and} \; w,$$

then $\qquad\qquad \tan 7\theta = 0,$

i.e. by Art. 30,

$$7t - {}^7C_3 \cdot t^3 + {}^7C_5 \cdot t^5 - {}^7C_7 t^7 = 0,$$

or $\qquad\qquad t^7 - 21t^5 + 35t^3 - 7t = 0,$

i.e. $\qquad\qquad t\{t^6 - 21t^4 + 35t^2 - 7\} = 0 \qquad\qquad \text{...(3)}.$

But

$$\tan \pi = 0, \; \tan\frac{6\pi}{7} = -\tan\frac{\pi}{7}, \; \tan\frac{5\pi}{7} = -\tan\frac{2\pi}{7} \; \text{and} \; \tan\frac{4\pi}{7} = -\tan\frac{3\pi}{7}.$$

The roots of (3) are therefore

$$0, \; \pm\tan\frac{\pi}{7}, \; \pm\tan\frac{3\pi}{7} \; \text{and} \; \pm\tan\frac{5\pi}{7}.$$

Hence, putting $t^2 = z$, the quantities

$$\tan^2\frac{\pi}{7}, \; \tan^2\frac{3\pi}{7}, \; \text{and} \; \tan^2\frac{5\pi}{7}$$

are the roots of (2).

EXAMPLES VI

1. Prove that

$$\left(x - 2\cos\frac{2\pi}{5}\right)\left(x - 2\cos\frac{4\pi}{5}\right)\left(x - 2\cos\frac{6\pi}{5}\right)\left(x - 2\cos\frac{8\pi}{5}\right)$$

$$= x^4 + 2x^3 - x^2 - 2x + 1.$$

2. Prove that the roots of the equation

$$8x^3 + 4x^2 - 4x - 1 = 0 \text{ are } \cos\frac{2\pi}{7}, \cos\frac{4\pi}{7} \text{ and } \cos\frac{6\pi}{7}.$$

3. Prove that $\sin\frac{2\pi}{7}, \sin\frac{4\pi}{7}$ and $\sin\frac{8\pi}{7}$ are the roots of the equation

$$z^3 - \frac{\sqrt{7}}{2}x^2 + \frac{\sqrt{7}}{8} = 0.$$

Prove that

4. $\dfrac{1}{4 - \sec^2\dfrac{2\pi}{7}} + \dfrac{1}{4 - \sec^2\dfrac{4\pi}{7}} + \dfrac{1}{4 - \sec^2\dfrac{6\pi}{7}} = 1$.

5. $\cos^4\dfrac{\pi}{9} + \cos^4\dfrac{2\pi}{9} + \cos^4\dfrac{3\pi}{9} + \cos^4\dfrac{4\pi}{9} = \dfrac{19}{16}$.

6. $\sec^4\dfrac{\pi}{9} + \sec^4\dfrac{2\pi}{9} + \sec^4\dfrac{3\pi}{9} + \sec^4\dfrac{4\pi}{9} = 1120$.

7. $\cos\dfrac{\pi}{11} + \cos\dfrac{3\pi}{11} + \cos\dfrac{5\pi}{11} + \cos\dfrac{7\pi}{11} + \cos\dfrac{9\pi}{11} = \dfrac{1}{2}$.

8. Form the equation whose roots are

$$\tan^2\frac{\pi}{11}, \ \tan^2\frac{2\pi}{11}, \ \tan^2\frac{3\pi}{11}, \ \tan^2\frac{4\pi}{11} \text{ and } \tan^2\frac{5\pi}{11}.$$

[Commence with equation (3) of Art. 30.]

Prove that

9. $\cot^2\dfrac{\pi}{11} + \cot^2\dfrac{2\pi}{11} + \cot^2\dfrac{3\pi}{11} + \cot^2\dfrac{4\pi}{11} + \cot^2\dfrac{5\pi}{11} = 15$.

10. $\sec^2\dfrac{\pi}{11} + \sec^2\dfrac{2\pi}{11} + \sec^2\dfrac{3\pi}{11} + \sec^2\dfrac{4\pi}{11} + \sec^2\dfrac{5\pi}{11} = 60$.

11. $\cos\dfrac{2\pi}{13} + \cos\dfrac{6\pi}{13} + \cos\dfrac{18\pi}{13} = \dfrac{-\sqrt{13} - 1}{4}$.

12. $\cos\dfrac{10\pi}{13} + \cos\dfrac{14\pi}{13} + \cos\dfrac{22\pi}{13} = \dfrac{-\sqrt{13} - 1}{4}$.

13. $\cos\dfrac{\pi}{15} + \cos\dfrac{7\pi}{15} + \cos\dfrac{11\pi}{15} + \cos\dfrac{13\pi}{15} = -\dfrac{1}{2}$.

14. Prove that $\sin\dfrac{\pi}{14}$ is a root of the equation

$$64x^6 - 80x^4 + 24x^2 - 1 = 0.$$

Expansions of Sines and Cosines of Multiple Angles, and of Powers of Sines and Cosines

[On a first reading of the subject the student is recommended to omit from the beginning of Art. 48 to the end of the chapter.]

■ **42.** In this chapter we shall show how to expand powers of cosines and sines of an angle in terms of cosines and sines of multiples of that angle, and also how to express cosines and sines of multiple angles in terms of powers of cosines and sines.

Throughout the chapter n denotes a positive integer.

■ **43.** Let $x \equiv \cos\theta + i\sin\theta$, so that

$$\frac{1}{x} = \frac{1}{\cos\theta + i\sin\theta} = \frac{\cos\theta - i\sin\theta}{\cos^2\theta + \sin^2\theta} = \cos\theta - i\sin\theta.$$

Hence,
$$x + \frac{1}{x} = 2\cos\theta,$$

and
$$x - \frac{1}{x} = 2i\sin\theta.$$

Also, by De Moivre's Theorem, we have
$$x^n = \cos n\theta + i\sin n\theta,$$

and
$$\frac{1}{x^n} = \cos n\theta - i\sin n\theta,$$

so that
$$x^n + \frac{1}{x^n} = 2\cos n\theta,$$

and
$$x^n - \frac{1}{x^n} = 2i\sin n\theta.$$

■ **44.** *To expand $\cos^n\theta$ in a series of cosines of multiples of θ, n being a positive integer.*

From the previous article, we have

$$(2\cos\theta)^n = \left(x + \frac{1}{x}\right)^n$$

$$= x^n + nx^{n-1} \cdot \frac{1}{x} + \frac{n(n-1)}{1\cdot 2}x^{n-2}\cdot\frac{1}{x^2} + \dots$$

43

$$+ \frac{n(n-1)}{1\cdot 2}x^2 \cdot \frac{1}{x^{n-2}} + nx \cdot \frac{1}{x^{n-1}} + \frac{1}{x^n}$$

$$= x^n + nx^{n-2} + \frac{n(n-1)}{1\cdot 2}x^{n-4} + \dots$$

$$+ \frac{n(n-1)}{1\cdot 2}\frac{1}{x^{n-4}} + n\cdot\frac{1}{x^{n-2}} + \frac{1}{x^n} \qquad \dots(1)$$

Taking together the first and last of these terms, the second and next to last, and so on, we have

$$(2\cos\theta)^n = \left(x^n + \frac{1}{x^n}\right) + n\left(x^{n-2} + \frac{1}{x^{n-2}}\right)$$

$$+ \frac{n(n-1)}{1\cdot 2}\left(x^{n-4} + \frac{1}{x^{n-4}}\right) + \dots$$

But by the last article we have

$$x^n + \frac{1}{x^n} = 2\cos n\theta, \quad x^{n-2} + \frac{1}{x^{n-2}} = 2\cos(n-2)\theta, \dots$$

Hence

$$2^n \cos^n\theta = 2\cos n\theta + n\cdot 2\cos(n-2)\theta + \frac{n(n-1)}{1\cdot 2}\cdot 2\cos(n-4)\theta + \dots,$$

i.e., $\qquad 2^{n-1}\cos^n\theta = \cos n\theta + n\cos(n-2)\theta + \frac{n(n-1)}{1\cdot 2}\cos(n-4)\theta + \dots \qquad \dots(2)$

If n be odd, there are an even number of terms on the right-hand side of (1), so that the terms take together in pairs and the last term contains $\cos\theta$.

If n be even, there are an odd number of terms on the right-hand side of (1), so that after all the possible pairs have been taken there is a term left not containing x. This term will, when divided by 2, form the last term on the right-hand of (2).

It could easily be shown that the last term is $\dfrac{\underline{|n}}{\underline{\left|\frac{n-1}{2}\right.}\,\underline{\left|\frac{n+1}{2}\right.}}\cos\theta$ if n be odd, and

$\dfrac{1}{2}\dfrac{\underline{|n}}{\left\{\underline{\left|\frac{n}{2}\right.}\right\}^2}$ if n be even.

■ **45. EXAMPLE 1.** *Expand $\cos^8\theta$ in a series of cosines of multiples of θ.*

We have $\qquad (2\cos\theta)^8 = \left(x + \dfrac{1}{x}\right)^8$

$$= x^8 + 8x^6 + 28x^4 + 56x^3 + 70 + 56\cdot\frac{1}{x^2} + 28\cdot\frac{1}{x^4} + 8\cdot\frac{1}{x^6} + \frac{1}{x^8}$$

$$= \left(x^8 + \frac{1}{x^8}\right) + 8\left(x^6 + \frac{1}{x^6}\right) + 28\left(x^4 + \frac{1}{x^4}\right) + 56\left(x^2 + \frac{1}{x^2}\right) + 70$$

$$= 2 \cdot \cos 8\theta + 8 \cdot 2 \cos 6\theta + 28 \cdot 2 \cos 4\theta + 56 \cdot 2 \cos 2\theta + 70,$$

$$\therefore \qquad 2^7 \cos^8 \theta = \cos 8\theta + 8 \cos 6\theta + 28 \cos 4\theta + 56 \cos 2\theta + 35.$$

▌ **EXAMPLE 2.** *Expand* $\cos^7 \theta$ *in a series of cosines of multiples of* θ.

We have $\qquad (2\cos\theta)^7 = \left(x + \frac{1}{x}\right)^7.$

$$= x^7 + 7 \cdot x^5 + 21x^3 + 35x + 35 \cdot \frac{1}{x} + 21 \cdot \frac{1}{x^3} + 7 \cdot \frac{1}{x^5} + \frac{1}{x^7}$$

$$= \left(x^7 + \frac{1}{x^7}\right) + 7\left(x^5 + \frac{1}{x^5}\right) + 21\left(x^3 + \frac{1}{x^3}\right) + 35\left(x + \frac{1}{x}\right)$$

$$= 2 \cdot \cos 7\theta + 7 \cdot 2 \cos 5\theta + 21 \cdot 2 \cos 3\theta + 35 \cdot 2 \cos\theta,$$

$$\therefore \qquad 2^6 \cos^7 \theta = \cos 7\theta + 7 \cos 5\theta + 21 \cos 3\theta + 35 \cos \theta.$$

▨ **46.** *To express* $\sin^n \theta$ *in a series of cosines or sines of multiples of* θ *according as* n *is an even or odd integer.*

By Art. 43 we have

$$2i \sin \theta = x - \frac{1}{x},$$

so that $\qquad 2^n i^n \sin^n \theta = \left(x - \frac{1}{x}\right)^n$ ...(1)

Case I. Let n be **even**, so that the last term in the expansion is

$$+\frac{1}{x^n}, \text{ and } i^n = (-1)^{n/2}.$$

The equation (1) is therefore

$$2^n (-1)^{n/2} \sin^n \theta = x^n - nx^{n-1} \cdot \frac{1}{x} + \frac{n(n-1)}{1 \cdot 2} x^{n-2} \cdot \frac{1}{x^2} - \cdots\cdots$$

$$+ \frac{n(n-1)}{1 \cdot 2} x^2 \cdot \frac{1}{x^{n-2}} - nx \cdot \frac{1}{x^{n-1}} + \frac{1}{x^n} \qquad \text{...(2)}$$

$$= \left(x^n + \frac{1}{x^n}\right) - n\left(x^{n-2} + \frac{1}{x^{n-2}}\right) + \frac{n(n-1)}{1 \cdot 2}\left(x^{n-4} + \frac{1}{x^{n-4}}\right) - \cdots\cdots,$$

$$= 2 \cdot \cos n\theta - n \cdot 2 \cos (n-2)\theta + \frac{n(n-1)}{1 \cdot 2} \cdot 2 \cos (n-4)\theta - \cdots\cdots,$$

as in Art. 44.

$$\therefore \quad 2^{n-1} (-1)^{n/2} \sin^n \theta = \cos n\theta - n \cos (n-2)\theta + \frac{n(n-1)}{1 \cdot 2} \cos (n-4)\theta - \cdots\cdots \qquad \text{...(3)}$$

Since n is even, there are an odd number of terms in (2), so that there will be a middle term which does not contain x. This term, on being divided by 2, will be the last term in equation (3).

This last term could easily be shown to be $\dfrac{1}{2}(-1)^{n/2}\dfrac{\lfloor n}{\left\{\left\lfloor\dfrac{n}{2}\right\rfloor\right\}^2}$.

Case II. Let n be **odd**, so that the last term in the expansion (1) will be

$$-\frac{1}{x^n}, \text{ and } i^n = i\cdot i^{n-1} = i(-1)\frac{n-1}{2}.$$

The equation (1) then becomes

$$2^n\cdot i\cdot(-1)^{n-1/2}\cdot\sin^n\theta = x^n - nx^{n-1}\cdot\frac{1}{x} + \frac{n(n-1)}{1\cdot 2}x^{n-2}\cdot\frac{1}{x^2}$$

$$\dots\dots - \frac{n(n-1)}{1\cdot 2}x^3\cdot\frac{1}{x^{n-2}} + nx\cdot\frac{1}{x^{n-1}} - \frac{1}{x^n}$$

$$= \left(x^n - \frac{1}{x^n}\right) - n\left(x^{n-2} - \frac{1}{x^{n-2}}\right) + \frac{n(n-1)}{1\cdot 2}\left(x^{n-4} - \frac{1}{x^{n-4}}\right)\dots \qquad \dots(4)$$

Now, by Art. 43, $x^n - \dfrac{1}{x^n} = 2i\sin n\theta$,

$$x^{n-2} - \frac{1}{x^{n-2}} = 2i\sin(n-2)\theta,$$

. . . .

Hence (4) becomes

$$2^n\cdot i\cdot(-1)^{-1/2}\sin^n\theta = 2i\sin n\theta - n\cdot 2i\sin(n-2)\theta$$

$$+\frac{n(n-1)}{1\cdot 2}\cdot 2i\sin(n-4)\theta - \dots,$$

so that $2^{n-1}(-1)^{-1/2}\sin^n\theta$

$$= \sin n\theta - n\sin(n-2)\theta + \frac{n(n-1)}{1\cdot 2}\sin(n-4)\theta - \dots\dots \qquad \dots(5)$$

Since n is in this case odd, there are an even number of terms in (4), so that (4) can be divided into pairs of terms, and there is no middle term. The last term in (5) therefore contains $\sin\theta$.

This last term could easily be shown to be $(-1)^{n-1/2}\dfrac{\lfloor n}{\dfrac{\lfloor n-1}{2}\dfrac{\lfloor n+1}{2}}\sin\theta$.

■ **47. EXAMPLE 1.** *Expand $\sin^6\theta$ in a series of cosines of multiples of θ.*

We have
$$2^6 i^6 \sin^6 \theta = \left(x - \frac{1}{x}\right)^6$$

$$= x^6 - 6x^4 + 15x^2 - 20 + 15 \cdot \frac{1}{x^2} - 6 \cdot \frac{1}{x^4} + \frac{1}{x^6};$$

so that
$$-2^6 \sin^6 \theta = \left(x^6 + \frac{1}{x^6}\right) - 6\left(x^4 + \frac{1}{x^4}\right) + 15\left(x^2 + \frac{1}{x^2}\right) - 20$$

$$= 2\cos 6\theta - 6 \cdot 2\cos 4\theta + 15 \cdot 2\cos 2\theta - 20.$$

$$\therefore \ -2^5 \sin^6 \theta = \cos 6\theta - 6\cos 4\theta + 15\cos 2\theta - 10.$$

▌ EXAMPLE 2. *Expand* $\sin^7 \theta$ *in a series of sines of multiples of* θ.

We have
$$2^7 i^7 \sin^7 \theta = \left(x - \frac{1}{x}\right)^7$$

$$= x^7 - 7x^5 + 21x^3 - 35x + 35 \cdot \frac{1}{x} - 21 \cdot \frac{1}{x^3} + 7 \cdot \frac{1}{x^5} - \frac{1}{x^7}$$

$$= \left(x^7 - \frac{1}{x^7}\right) - 7\left(x^5 - \frac{1}{x^5}\right) + 21\left(x^3 - \frac{1}{x^3}\right) - 35\left(x - \frac{1}{x}\right).$$

$$\therefore \ -2^7 \cdot i \cdot \sin^7\theta = 2i\sin 7\theta - 7 \cdot 2i\sin 5\theta + 21 \cdot 2i\sin 3\theta - 35 \cdot 2i\sin \theta.$$

$$\therefore \ -2^6 \sin^7\theta = \sin 7\theta - 7\sin 5\theta + 21\sin 3\theta - 35\sin \theta.$$

▌ EXAMPLE 3. *Expand* $\cos^5 \theta \sin^7 \theta$ *in a series of sines of multiples of* θ.
We have

$$2^5 \cos^5 \theta = \left(x + \frac{1}{x}\right)^5, \text{ and } 2^7 i^7 \sin^7 \theta = \left(x - \frac{1}{x}\right)^7.$$

Hence,
$$2^{12} \cdot i^7 \cdot \cos^5 \theta \sin^7 \theta = \left(x^2 - \frac{1}{x^2}\right)^5 \left(x - \frac{1}{x}\right)^2$$

$$= \left[x^{10} - 5x^6 + 10x^2 - \frac{10}{x^2} + \frac{5}{x^6} - \frac{1}{x^{10}}\right]\left[x^3 - 2 + \frac{1}{x^3}\right]$$

$$= \left(x^{12} - \frac{1}{x^{12}}\right) - 2\left(x^{10} - \frac{1}{x^{10}}\right) - 4\left(x^8 - \frac{1}{x^8}\right) + 10\left(x^6 - \frac{1}{x^6}\right)$$

$$. +5\left(x^4 - \frac{1}{x^4}\right) - 20\left(x^2 - \frac{1}{x^2}\right).$$

Hence, as before, we have

$$-2^{11} \cos^5\theta \sin^7\theta = \sin 12\theta - 2\sin 10\theta - 4\sin 8\theta + 10\sin 6\theta + 5\sin 4\theta$$
$$- 20\sin 2\theta.$$

EXAMPLES VII

Prove that

1. $\sin^5 \theta = \dfrac{1}{16}$ [sin 5 θ – 5 sin 3 θ + 10 sin θ].

2. $\cos^9 \theta = \dfrac{1}{256}$ [cos 9 θ + 9 cos 7 θ + 36 cos 5 θ + 84 cos 3 θ + 126 cos θ].

3. $\cos^{10} \theta = \dfrac{1}{512}$ [cos 10 θ + 10 cos 8 θ + 45 cos 6 θ + 120 cos 4 θ + 210 cos 2 θ + 126].

4. $\sin^8 \theta = \dfrac{1}{128}$ [cos 8 θ – 8 cos 6 θ + 28 cos 4 θ – 56 cos 2 θ + 35].

5. $\sin^9 \theta = \dfrac{1}{256}$ [sin 9 θ – 9 sin 7 θ + 36 sin 5 θ – 84 sin 3 θ + 126 sin θ].

6. $2^5 \sin^4 \theta \cos^2 \theta = \cos 6 \theta - 2 \cos 4 \theta - \cos 2 \theta + 2$.

7. $2^6 \sin^5 \theta \cos^2 \theta = \sin 7 \theta - 3 \sin 5 \theta + \sin 3 \theta + 5 \sin \theta$.

8. $-2^{10} \sin^3 \theta \cos^8 \theta = \sin 11 \theta + 5 \sin 9 \theta + 7 \sin 7 \theta - 5 \sin 5 \theta - 22 \sin 3 \theta - 14 \sin \theta$.

■ ****48. *To express*** $\dfrac{\sin n\theta}{\sin \theta}$ *in a series of descending powers of cos* θ.

If x be < 1, we have

$$\frac{\sin \theta}{1 - 2x \cos \theta + x^2} = \sin \theta + x \sin 2\theta + x^2 \sin 3\theta + \cdots$$

$$+ x^{n-1} \sin n\theta + \dots \text{ad inf.} \qquad \dots(1)$$

This may be shown by multiplying each side by
$$1 - 2x \cos \theta + x^2,$$
when it will be found that the right-hand member will reduce to sin θ.

A more rigorous proof will be found in Chap. VIII.

Equating coefficients of x^{n-1} in (1), we have

$$\frac{\sin n\theta}{\sin \theta} = \text{coefficient of } x^{n-1} \text{ in } \left[1 - 2x \cos\theta + x^2\right]^{-1}$$

= coefficient of x^{n-1} in $[1 - x(2 \cos \theta - x)]^{-1}$

= coefficient of x^{n-1} in

$$1 + x(2\cos\theta - x) + x^2 (2\cos\theta - x)^2 + \cdots\cdots$$

$$+ x^{n-6} (2\cos\theta - x)^{n-3} + x^{n-2} (2\cos\theta - x)^{n-2}$$

$$+ x^{n-1} (2\cos\theta - x)^{n-1} + x^n (2\cos\theta - x)^n + \cdots\cdots \qquad \dots(2)$$

Now coefficient of
$$x^{n-1} \text{ in } x^{n-1} (2 \cos \theta - x)^{n-1} = (2 \cos \theta)^{n-1},$$

coefficient of x^{n-1} in $x^{n-2}(2\cos\theta - x)^{n-2}$

$\quad$ = coefficient of x in $(2\cos\theta - x)^{n-2}$

$\quad$ = $-(n-2)(2\cos\theta)^{n-3}$,

coefficient of x^{n-1} in $x^{n-3}(2\cos\theta - x)^{n-3}$

$\quad$ = coefficient of x^2 in $(2\cos\theta - x)^{n-3}$

$\quad = \dfrac{(n-3)(n-4)}{1\cdot 2}(2\cos\theta)^{n-5}$,

and so on.

Hence, from (2), picking out in this manner all the coefficients of x^{n-1}, we have

$$\frac{\sin n\theta}{\sin\theta} = (2\cos\theta)^{n-1} - (n-2)(2\cos\theta)^{n-6}$$

$$+\frac{(n-3)(n-4)}{1\cdot 2}(2\cos\theta)^{n-5}$$

$$-\frac{(n-4)(n-5)(n-6)}{1\cdot 2\cdot 3}(2\cos\theta)^{n-7} +\cdots\cdots \qquad\qquad \cdots$$

If n be odd, the last term could be proved to be $(-1)^{\frac{n-1}{2}}$; if n be even, it could be

shown to be $(-1)^{\frac{n}{2}-1}(n\cos\theta)$.

▨ ****49.** *To express* cos $n\theta$ *in a series of descending powers of* cos θ.

If x be < 1, we have

$$\frac{1-x^2}{1-2x\cos\theta + x^2} = 1 + 2x\cos\theta + 2x^2\cos 2\theta + 2x^3\cos 3\theta +\cdots$$

$$\cdots + 2x^n\cos n\theta +\ldots\text{ad inf.} \qquad\qquad\ldots(1).$$

This may be shown by multiplying both the sides by

$$1 - 2x\cos\theta + x^2,$$

when it will be found that all the terms on the right-hand side will reduce to $1-x^2$.

A more rigorous proof will be found in Chap. VIII.

Equating coefficients of x^n on the two sides of (1), we have

$2\cos n\theta$ = coefficient of x^n in $(1-x^2)[1-2x\cos\theta + x^2]^{-1}$

$\quad$ = coefficient of x^n − coefficient of x^{n-2} in $[1 - x(2\cos\theta - x)]^{-1}$

$\quad$ = coefficient of x^n − coefficient of x^{n-2} in

$$1 + x(2\cos\theta - x) + x^2(2\cos\theta - x)^2 +\cdots$$

$$\cdots + x^{n-2}(2\cos\theta - x)^{n-2} + x^{n-1}(2\cos\theta - x)^{n-1}$$

$$+ x^n(2\cos\theta - x)^n + x^{n+1}(2\cos\theta - x)^{n+1} +\cdots.$$

Picking out the required coefficients as in the last article, starting with the term

$$x^n (2 \cos \theta - x)^n,$$

we have $2 \cos n\theta$

$$= (2\cos\theta)^n - (n-1)(2\cos\theta)^{n-2} + \frac{(n-2)(n-3)}{1\cdot 2}(2\cos\theta)^{n-1}$$

$$- \frac{(n-3)(n-4)(n-5)}{1\cdot 2\cdot 3}(2\cos\theta)^{n-6} + \cdots\cdots$$

$$- \left[(2\cos\theta)^{n-2} - (n-3)(2\cos\theta)^{n-4} + \frac{(n-4)(n-5)}{1\cdot 2}(2\cos\theta)^{n-6} - \cdots\cdots\right]$$

$$= (2\cos\theta)^n - n(2\cos\theta)^{n-2} + \left[\frac{(n-2)(n-3)}{1\cdot 2} + (n-3)\right](2\cos\theta)^{n-4}$$

$$- \left[\frac{(n-3)(n-4)(n-5)}{1\cdot 2\cdot 3} + \frac{(n-4)(n-5)}{1\cdot 2}\right](2\cos\theta)^{n-6} + \ldots,$$

so that, finally,

$$2\cos n\theta = (2\cos\theta)^n - n(2\cos\theta)^{n-2} + \frac{n(n-3)}{1\cdot 2}(2\cos\theta)^{n-4}$$

$$- \frac{n(n-4)(n-5)}{1\cdot 2\cdot 3}(2\cos\theta)^{n-6} + \cdots\cdots \qquad \ldots(2).$$

The last term could be shown to be

$$(-1)^{\frac{n-1}{2}} \cdot n \cdot (2\cos\theta) \text{ or } (-1)^{\frac{n}{3}}. 2,$$

according as n is odd or even.

▮ ** **50.** *To expand* $\dfrac{\sin n\theta}{\sin \theta}$ *in a series of ascending powers of cos* θ.

As in Art. 48, we have

$$\frac{\sin n\theta}{\sin \theta} = \text{coefficient of } x^{n-1} \text{ in } [1 - 2x \cos \theta + x^2]^{-1}$$

$$= \text{coefficient of } x^{n-1} \text{ in } [1 + x(x - 2\cos\theta)]^{-1}$$

$$= \text{coefficient of } x^{n-1} \text{ in}$$

$$1 - x (x - 2\cos\theta) + x^2(x - 2\cos\theta)^2 - \cdots\cdots$$

$$\cdots\cdots + (-1)^r x^r (x - 2\cos\theta)^r + \cdots \qquad \ldots(1).$$

Case I. Let n be **odd**, so that $(n - 1)$ is even.

The lowest term in (1) which gives any coefficient of x^{n-1} is then that for which

$$r = \frac{n-1}{2}.$$

Hence, in this case,

$$\frac{\sin n\theta}{\sin \theta} = \text{coefficient of } x^{n-1} \text{ in } 1 - x(x - 2\cos\theta) + \dots$$

$$+ (-1)^{\frac{n-1}{2}} \cdot x^{\frac{n-1}{2}} (x - 2\cos\theta)^{\frac{n-1}{2}} + (-1)^{\frac{n+1}{2}} \cdot x^{\frac{n+1}{2}} (x - 2\cos\theta)^{\frac{n+1}{2}}$$

$$+ (-1)^{\frac{n+3}{2}} x^{\frac{n+3}{2}} (x - 2\cos\theta)^{\frac{n+3}{2}} + \dots\dots + (-1)^{n-1} x^{n-1} (x - 2\cos\theta)^{n-1} + \dots\dots$$

Picking out the required coefficients as in Art. 48, we have

$$\frac{\sin n\theta}{\sin \theta} = (-1)^{\frac{n-1}{2}} + (-1)^{\frac{n+1}{2}} \left[\frac{\frac{n+1}{2} \cdot \frac{n-1}{2}}{1 \cdot 2} \right] (-2\cos\theta)^2$$

$$+ (-1)^{\frac{n+3}{2}} \cdot \frac{\frac{n+3}{2} \cdot \frac{n+1}{2} \cdot \frac{n-1}{2} \cdot \frac{n-3}{2}}{1 \cdot 2 \cdot 3 \cdot 4} (-2\cos\theta)^4 + \dots + (2\cos\theta)^{n-1}.$$

Hence, finally, when n is **odd**, we have

$$(-1)^{\frac{n-1}{2}} \cdot \frac{\sin n\theta}{\sin \theta} = 1 - \frac{n^2 - 1^2}{1 \cdot 2} \cos^2\theta + \frac{(n^2 - 1^2)(n^2 - 3^2)}{\underline{|4}} \cos^4\theta$$

$$- \frac{(n^2 - 1^2)(n^2 - 3^2)(n^2 - 5^2)}{\underline{|6}} \cos^6\theta - \dots\dots$$

$$+ (-1)^{\frac{n-1}{2}} (2\cos\theta)^{n-1} \dots\dots \qquad\qquad \dots(2).$$

Case II. Let n be **even**, so that $n - 1$ is odd.

The lowest term in (1) which gives any coefficient of x^{n-1} is then that for which

$$r = \frac{n}{2}.$$

Hence, in this case,

$$\frac{\sin n\theta}{\sin \theta} = \text{coefficient of } x^{n-1} \text{ in } 1 - x(x - 2\cos\theta) + \dots$$

$$+ (-1)^{\frac{n}{2}} x^{\frac{n}{2}} (x - 2\cos\theta)^{\frac{n}{2}} + (-1)^{\frac{n}{2}+1} x^{\frac{n}{2}+1} (x - 2\cos\theta)^{\frac{n}{2}+1}$$

$$+ (-1)^{\frac{n}{2}+2} x^{\frac{n}{2}+2} (x - 2\cos\theta)^{\frac{n}{2}+2} + \dots + (-1)^{n-1} x^{n-1} (x - 2\cos\theta)^{n-1} +$$

Picking out the required coefficients, we have

$$\frac{\sin n\theta}{\sin \theta} = (-1)^{\frac{n}{2}} \cdot \frac{n}{2} (-2\cos\theta)$$

$$+ (-1)^{\frac{n}{2}+1} \frac{\left(\frac{n}{2}+1\right)\left(\frac{n}{2}\right)\left(\frac{n}{2}-1\right)}{1 \cdot 2 \cdot 3} (-2\cos\theta)^3$$

$$+(-1)^{\frac{n}{2}+2} \cdot \frac{\left(\frac{n}{2}+2\right)\left(\frac{n}{2}+1\right)\frac{n}{2}\left(\frac{n}{2}-1\right)\left(\frac{n}{2}-2\right)}{1\cdot 2\cdot 3\cdot 4\cdot 5}(-2\cos\theta)^5$$

$$+ \;.................. + (2\cos\theta)^{n-1}.$$

Hence, finally, when n is **even**, we have

$$(-1)^{\frac{n}{2}+1}\frac{\sin n\theta}{\sin\theta}$$

$$= n\cos\theta - \frac{n(n^2-2^2)}{\underline{|3}}\cos^3\theta + \frac{n(n^2-2^2)(n^2-4^2)}{\underline{|5}}\cos^5\theta$$

$$-\;...... + (-1)^{\frac{n}{2}+1}(2\cos\theta)^{n-1} \qquad\qquad ...(3)$$

N.B. It will be noted that equations (2) and (3) of this article are simply the series of Art. 48 written backwards. This is clear from the method of proof, or the statement could be easily verified independently.

■ ****51.** *To expand* $\cos n\theta$ *in a series of ascending powers of* $\cos\theta$.

As in Art. 49, we have

$2\cos n\theta =$ coefficient of x^n – coefficient of x^{n-2} in $(1-2x\cos\theta+x^2)^{-1}$.

$\qquad\qquad$ = coefficient of x^n – coefficient of x^{n-2} in

$\qquad 1-x(x-2\cos\theta)+x^2(x-2\cos\theta)^2 - ...$

$$\qquad\qquad + (-1)^r x^r (x-2\cos\theta)^r + ... \qquad\qquad ...(1),$$

as in Art. 49.

Case I. Let n be **odd**, so that $n-1$ is even.

The lowest term in (1) which will give any of the coefficients we want is that for which

$$r = \frac{n-1}{2}.$$

Hence, $2\cos n\theta =$ coefficient of x^n – coefficient of x^{n-1} in

$$1-x(x-2\cos\theta)+\cdots+(-1)^{\frac{n-1}{2}} x^{\frac{n-1}{2}} (x-2\cos\theta)^{\frac{n-1}{2}}$$

$$+(-1)^{\frac{n+1}{2}} x^{\frac{n-1}{2}} (x-2\cos\theta)^{\frac{n+1}{2}} +(-1)^{\frac{n+3}{2}} x^{\frac{n+3}{2}} (x-2\cos\theta)^{\frac{n+3}{2}}$$

$$+ \cdots + (-1)^n x^n (x-2\cos\theta)^n \cdots$$

$$= (-1)^{\frac{n-1}{2}}\left[-\frac{n-1}{2}(-2\cos\theta)\right]$$

$$+(-1)^{\frac{n-1}{2}}\left[\frac{n+1}{2}(-2\cos\theta) - \frac{\frac{n+1}{2}\cdot\frac{n-1}{2}\cdot\frac{n-3}{2}}{1\cdot 2\cdot 3}(-2\cos\theta)^3\right]$$

$$+(-1)^{\frac{n+3}{2}}\left[\frac{\frac{n+3}{2}\cdot\frac{n+1}{2}\cdot\frac{n-1}{2}}{1\cdot2\cdot3}(-2\cos\theta)^3\right.$$

$$\left.-\frac{\frac{n+3}{2}\frac{n+1}{2}\frac{n-1}{2}\frac{n-3}{2}\frac{n-5}{2}}{1\cdot2\cdot3\cdot4\cdot5}(-2\cos\theta)^5\right]$$

$$+\ldots\ldots\ldots\ldots+(2\cos\theta)^n.$$

$$\therefore\quad(-1)^{\frac{n-1}{2}}\cdot2\cos n\theta$$

$$=\cos\theta[(n-1)+(n+1)]-\frac{(n+1)(n-1)}{\underline{3}}\cos^3\theta[(n-3)+(n+3)]$$

$$+\frac{(n+3)(n+1)(n-1)(n-3)}{\underline{5}}\cos^5\theta[(n-5)+(n+5)]+\ldots$$

$$+(-1)^{\frac{n-1}{2}}(2\cos\theta)^n$$

Hence, finally, when n is **odd,**

$$(-1)^{\frac{n-1}{2}}\cos n\theta$$

$$=n\cos\theta-\frac{n\left(n^2-1^2\right)}{\underline{3}}\cos^3\theta+\frac{n\left(n^2-1^2\right)\left(n^2-3^2\right)}{\underline{5}}\cos^5\theta$$

$$-\ldots(-1)^{\frac{n-1}{2}}\cdot2^{n-1}\cos^n\theta\qquad\ldots(2)$$

Case II. Let n be **even.**

The lowest term in (1) which will give any of the required coefficients in that for which

$$r=\frac{n-2}{2}.$$

Hence we have

$2\cos n\theta$ = coefficient of x^n – coefficient of x^{n-2} in

$$1-x(x-2\cos\theta)+\cdots+(-1)^{\frac{n-2}{2}}x^{\frac{n-2}{2}}(x-2\cos\theta)^{\frac{n-2}{2}}$$

$$+(-1)^{\frac{n}{2}}x^{\frac{n}{2}}(x-2\cos\theta)^{\frac{n}{2}}+(-1)^{\frac{n+2}{2}}x^{\frac{n+2}{2}}(x-2\cos\theta)^{\frac{n+2}{2}}$$

$$+\cdots+(-1)^n x^n(x-2\cos\theta)^n+\cdots$$

$$=(-1)^{\frac{n-2}{2}}[-1]+(-1)^{\frac{n}{2}}\left[1-\frac{\frac{n}{2}\cdot\frac{n-2}{2}}{1\cdot2}(-2\cos\theta)^2\right]$$

$$+(-1)^{\frac{n+2}{2}}\left[\frac{\frac{n+2}{2}\cdot\frac{n}{2}}{1\cdot 2}(-2\cos\theta)^2 \quad \frac{\frac{n+2}{2}\cdot\frac{n}{2}\cdot\frac{n-2}{2}\cdot\frac{n-4}{2}}{1\cdot 2\cdot 3\cdot 4}(-2\cos\theta)^4\right]$$

$$+(-1)^{\frac{n+4}{2}}\left[\frac{\frac{n+4}{2}\cdot\frac{n+2}{2}\cdot\frac{n}{2}\cdot\frac{n-2}{2}}{1\cdot 2\cdot 3\cdot 4}(-2\cos\theta)^4\right.$$

$$\left.-\frac{\frac{n+4}{2}\cdot\frac{n+2}{2}\cdot\frac{n}{2}\cdot\frac{n-2}{2}\cdot\frac{n-4}{2}\cdot\frac{n-6}{2}}{\underline{|6}}(-2\cos\theta)^6\right]$$

$$+\;\dots\dots\dots\dots\; +(2\cos\theta)^n.$$

$\therefore \qquad (-1)^{\frac{n}{2}}\cdot 2\cos n\theta$

$$=[1+1]-\frac{\cos^2\theta}{\underline{|2}}[n(n-2)+(n+2)\cdot n]$$

$$+\frac{\cos^4\theta}{\underline{|4}}[(n+2)\cdot n\cdot(n-2)(n-4)+(n+4)(n+2)\cdot n\cdot(n-2)]$$

$$+\dots\dots\dots+(-1)^{\frac{n}{2}}\cdot(2\cos\theta)^n.$$

Hence, finally, when n is **even**,

$$(-1)^{\frac{n}{2}}\cos n\,\theta=1-\frac{n^2\cos^2\theta}{\underline{|2}}+\frac{n^2\left(n^2-2^2\right)}{\underline{|4}}\cos^4\theta$$

$$-\frac{n^2\left(n^2-2^2\right)\left(n^2-4^2\right)}{\underline{|6}}\cos^6\theta+\dots$$

$$-(-1)^{\frac{n}{2}}2^{n-1}\cos^n\theta \qquad\qquad\dots(3)$$

N.B. As before, the equations (2) and (3) of this article are only the series (2) of Art. 49 written backwards.

▮ ****52.** From equation (2) of Art. 50 and equation (2) of Art. 51 we have, if n be **odd,**

$$(-1)^{\frac{n-1}{2}}\frac{\sin n\theta}{\sin\theta}=1-\frac{n^2-1^2}{\underline{|2}}\cos^2\theta+\frac{\left(n^2-1^2\right)\left(n^2-3^2\right)}{\underline{|4}}\cos^4\theta$$

$$-\frac{\left(n^2-1^2\right)\left(n^2-3^2\right)\left(n^2-5^2\right)}{\underline{|6}}\cos^6\theta+\dots\dots$$

$$+(-1)^{\frac{n-1}{2}}(2\cos\theta)^{n-1}\ldots\ldots \qquad \ldots(1)$$

and
$$(-1)^{-\frac{1}{2}}\cos n\theta = n\cos\theta - \frac{n(n^2-1^2)}{\underline{3}}\cos^2\theta$$

$$+\frac{(n^2-1^2)(n^2-3^2)}{\underline{5}}\cos^5\theta +\cdots\cdots +(-1)^{\frac{n-1}{2}}2^{n-1}\cos^n\theta \qquad \ldots(2)$$

In these equations change θ into $\frac{\pi}{2}-\theta$, and therefore $\cos\theta$ into $\sin\theta$.

Then $\sin n\theta$ will become

$$\sin\left(\frac{n\pi}{2}-n\theta\right), \text{ i.e., } (-1)^{\frac{n-1}{2}}\cos n\theta,$$

and $\cos n\theta$ will become

$$\cos\left(\frac{n\pi}{2}-n\theta\right), \text{ i.e., } (-1)^{\frac{n-1}{2}}\sin n\theta.$$

On making these substitutions we shall have, if n be **odd**,

$$\cos n\theta = \cos\theta\left\{1 - \frac{n^2-1^2}{\underline{2}}\sin^2\theta + \frac{(n^2-1^2)(n^2-3^2)}{\underline{4}}\sin^4\theta -\ldots\right.$$

$$\left.+(-1)^{\frac{n-1}{2}}\cdot 2^{n-1}\sin^{n-1}\theta\right\} \qquad \ldots(3),$$

and

$$\sin n\theta = n\sin\theta - \frac{n(n^2-1^2)}{\underline{3}}\sin^3\theta + \frac{n(n^2-1^2)(n^2-3^2)}{\underline{5}}\sin^5\theta$$

$$+\cdots\cdots +(-1)^{\frac{n-1}{2}}2^{n-1}\sin^n\theta \qquad \ldots(4).$$

■ ****53.** Again from equation (3) of Art. 50 and equation (3) of Art. 51 we have, if n be **even,**

$$(-1)^{\frac{n}{2}+1}\frac{\sin n\theta}{\sin\theta} = n\cos\theta - \frac{n(n^2-2^2)}{\underline{3}}\cos^3\theta$$

$$|\frac{n(n^2-2^2)(n^2-4^2)}{\underline{5}}\cos^5\theta \,|\ldots\ldots\,|\,(\,1)^{\frac{n}{2}+1}(2\cos\theta)^{n-1} \qquad \ldots(1),$$

and

$$(-1)^{\frac{n}{2}}\cos n\theta = 1 - \frac{n^2}{\underline{2}}\cos^2\theta + \frac{n^2(n^2-2^2)}{\underline{4}}\cos^4\theta -\ldots\ldots$$

$$+(-1)^{\frac{n}{2}}2^{n-1}(\cos^n\theta) \qquad \ldots(2).$$

In these equations change θ into $\frac{\pi}{2} - \theta$, and therefore $\cos \theta$ into $\sin \theta$.

Then $\sin n\theta$ will become

$$\sin\left(\frac{n\pi}{2} - n\theta\right), \ i.e., (-1)^{\frac{n}{2}+1} \sin n\theta,$$

and $\cos n\theta$ will become

$$\cos\left(\frac{n\pi}{2} - n\theta\right), \ i.e., (-1)^{\frac{n}{2}} \cos n\theta.$$

On making these substitutions we have, if n be **even,**

$$\frac{\sin n\theta}{\cos \theta} = n\sin \theta - \frac{n(n^2 - 2^2)}{\underline{|3}}\sin^3 \theta + \frac{n(n^2 - 2^2)(n^2 - 4^2)}{\underline{|5}}\sin^5 \theta \ldots$$

$$+ (-1)^{\frac{n}{2}+1}(2\sin\theta)^{n-1} \qquad \qquad \ldots(3),$$

and

$$\cos n\theta = 1 - \frac{n^2}{\underline{|2}}\sin^2 \theta + \frac{n^2(n^2 - 2^2)}{\underline{|4}}\sin^4 \theta$$

$$+ \ldots (-1)^{\frac{n}{2}} 2^{n-1}\sin^n \theta \qquad \qquad \ldots(4).$$

■ ****54.** Equations (1) and (2) of Art. 52 and equations (1) and (2) of Art. 53 give the expansions of $\sin n\theta$ and $\cos n\theta$ in ascending powers of $\cos \theta$ for the cases when n is even or odd. Equations (3) and (4) of the same two articles give the expansions of the same two quantities in terms of $\sin \theta$.

EXAMPLES VIII

1. $\sin 7\theta = 7 \sin \theta - 56 \sin^2 \theta + 112 \sin^5 \theta - 64 \sin^7 \theta.$
2. $\cos 7\theta = 64 \cos^7 \theta - 112 \cos^5 \theta + 56 \cos^3 \theta - 7 \cos \theta.$
3. $\sin 8\theta = \sin \theta \ [128 \cos^7 \theta - 192 \cos^3 \theta + 80 \cos^3 \theta - 8 \cos \theta].$
4. $\cos 8\theta = 1 - 32 \sin^2 \theta + 160 \sin^4 \theta - 256 \sin^6 \theta + 128 \sin^8 \theta.$
5. $\sin 9\theta = \sin \theta \ [256 \cos^8 \theta - 448 \cos^5 \theta + 240 \cos^4 \theta - 40 \cos^2 \theta + 1].$
6. Express $\cos 6\theta$ in terms of $\cos \theta$ only and verify for the cases

$$\theta = \frac{\pi}{3}, \theta = \frac{\pi}{2}$$

   respectively.
7. Prove the algebraic identity

$$p^n + q^n = (p + q)^n - n(p + q)^{n-2}pq + \frac{n(n-3)}{1 \cdot 2}(p + q)^{n-4} - p^2q^2 + \ldots$$

Deduce that

$$2 \cos n\theta = (2 \cos \theta)^n - n(2 \cos \theta)^{n-2} + \frac{n(n-3)}{1 \cdot 2}(2\cos\theta)^{n-4} - \dots.$$

▮ **55. EXAMPLE.** *Find the value of*

$$\sec \theta + \sec\left(\theta + \frac{2\pi}{n}\right) + \sec\left(\theta + \frac{4\pi}{n}\right) + \dots \text{ to } n \text{ terms.}$$

$$\sec^2 \theta + \sec^2\left(\theta + \frac{2\pi}{n}\right) + \sec^2\left(\theta + \frac{4\pi}{n}\right) + \dots \text{ to } n \text{ terms.}$$

From equations (2) and (3) of Art. 51, we know that

$$nc - \frac{n(n^2 - 1^2)}{\underline{3}}c^3 + \frac{n(n^2 - 1^2)(n^2 - 3^2)}{\underline{5}}c^5 + \dots + (-1)^{\frac{n-1}{2}} 2^{n-1} c^n$$

$$= (-1)^{\frac{n-1}{2}} \cdot \cos n\theta \qquad \dots(1),$$

when n is **odd**,

and that

$$1 - \frac{n^2}{\underline{2}}c^2 + \frac{n^2(n^3 - 2^3)}{\underline{4}}c^4 + \dots + (-1)^{\frac{n}{2}} 2^{n-1} c^n = (-1)^{\frac{n}{2}} \cos n\theta \qquad \dots(2),$$

when n is **even**,

where in each series c stands for $\cos \theta$.

If $\cos n\theta$ be now given, the equations (1) and (2) give $\cos \theta$.

But since $\cos n\theta = \cos(n\theta + 2\pi) = \cos(n\theta + 4\pi)$

$$= \dots\dots\dots\dots,$$

these equations would also give

$$\cos\left(\theta + \frac{2\pi}{n}\right), \cos\left(\theta + \frac{4\pi}{n}\right), \dots\dots$$

Hence, in each case, the roots are

$$\cos\theta, \cos\left(\theta + \frac{2\pi}{n}\right), \cos\left(\theta + \frac{4\pi}{n}\right), \dots\dots \text{ to } n \text{ terms.}$$

In (1) and (2) put $c = \dfrac{1}{y}$ and multiply by y^n.

We have then the equations

$$(-1)^{\frac{n-1}{2}} \cos n\theta \times y^n - n \cdot y^{n-1} + \frac{n(n^2 - 1^2)}{\underline{3}} y^{n-3} - \dots = 0 \qquad \dots(3),$$

when n is **odd**,

and

$$\left[(-1)^{\frac{n}{2}} \cos n\theta - 1\right] y^n + \frac{n^2}{\underline{2}} y^{n-3} - \dots = 0 \qquad \dots(4),$$

when n is **even**.

The roots of these equations are respectively

$$\sec\theta, \ \sec\left(\theta + \frac{2\pi}{n}\right), \ \sec\left(\theta + \frac{4\pi}{n}\right), \ ......$$

Call these $y_1, y_2, ..., y_n$.

Then

$$y_1 + y_2 + \cdots + y_n = \text{sum of the roots}$$

$$= \frac{n}{(-1)^{\frac{u\,|1}{2}}\cos n\theta} = (-1)^{\frac{n-1}{2}} n\sec n\theta, \text{ when } n \text{ is odd,}$$

and $= 0$, when n is even.

Also,

$$y_1^2 + y_2^2 + \cdots + y_n^2 = (y_1 + y_2 + \cdots + y_n)^2 - 2(y_1 y_2 + y_2 y_3 + ...)$$

$$= \frac{n^2}{\cos^2 n\theta} = n^2 \sec^2 n\theta, \text{ when } n \text{ is odd,}$$

and

$$= -2 \cdot \frac{\frac{n^2}{\lfloor 2}}{(-1)^{\frac{n}{2}}\cos n\theta - 1} = \frac{n^2}{1 - (-1)^{\frac{n}{2}}\cos n\theta}, \text{ when } n \text{ is even.}$$

EXAMPLES IX

Find the value of

1. $\cos\theta\cos\left(\theta + \frac{2\pi}{n}\right)\cos\left(\theta + \frac{4\pi}{n}\right)......\cos\left\{\theta + (n-1)\frac{2\pi}{n}\right\}$.

2. $\sin\theta\sin\left(\theta + \frac{2\pi}{n}\right)\sin\left(\theta + \frac{4\pi}{n}\right)......\sin\left\{\theta + (n-1)\frac{2\pi}{n}\right\}$.

3. $\operatorname{cosec}^2\theta + \operatorname{cosec}^2\left(\theta + \frac{2\pi}{n}\right) + \operatorname{cosec}^2\left(\theta + \frac{4\pi}{n}\right)$....... to n terms.

4. $\tan^2\theta + \tan^2\left(\theta + \frac{2\pi}{n}\right) + \tan^2\left(\theta + \frac{4\pi}{n}\right)$..... to n terms.

[*For the following 5 questions commence with equation (5) of Art. 80.*]

5. $\tan\theta + \tan\left(\theta + \frac{\pi}{n}\right) + \tan\left(\theta + \frac{2\pi}{n}\right)$...... to n terms.

6. $\cot\theta + \cot\left(\theta + \frac{\pi}{n}\right) + \cot\left(\theta + \frac{2\pi}{n}\right)$ .... to n terms.

7. $\tan\theta\tan\left(\theta + \frac{\pi}{n}\right)\tan\left(\theta + \frac{2\pi}{n}\right)$.....to n factors.

8. $\tan^2\theta + \tan^2\left(\theta + \dfrac{\pi}{n}\right) + \tan^2\left(\theta + \dfrac{2\pi}{n}\right) + \ldots\ldots$ to n terms.

9. If n be odd, prove that $S = 3C = n^2 - 1$, where

$$S = \sec^2\frac{\pi}{n} + \sec^2\frac{2\pi}{n} + \sec^2\frac{3\pi}{n} + \cdots\cdots \text{ to } n-1 \text{ terms,}$$

and $C = \csc^2\dfrac{\pi}{n} + \csc^2\dfrac{2\pi}{n} + \csc^2\dfrac{3\pi}{n} + \ldots\ldots$ to $n-1$ terms.

10. Find the sum of the products, taken two at a time, of expressions of the form

$\sec\left(\theta + \dfrac{2r\pi}{n}\right)$, where r has all values from zero to $n - 1$.

Note. When the student has become acquainted with the later chapters, he will find that the results of Arts. 49 and 51 are rather more easily obtained by starting with the expansion of Art. 110, so that $\dfrac{2\cos n\theta}{n}$

= coefficient of x^n in the expansion of $-\log\left[1 - x(2\cos\theta - x)\right]$.

Exponential Series for Complex Quantities. Circular Functions for Complex Angles. Hyperbolic Functions

■ **56.** When x is a real quantity we have proved in Art. 5 that

$$e^x = 1 + x + \frac{x^2}{\underline{|2}} + \frac{x^3}{\underline{|3}} + \text{ad inf} \qquad \qquad ...(1)$$

When x is not real but is complex, *i.e.*, of the form $a + b\sqrt{-1}$, the expression e^x has no meaning at present.

Let us so define it that for **all** values of x (whether real or complex) it shall mean the series.

$$1 + x + \frac{x^2}{\underline{|2}} + \frac{x^3}{\underline{|3}} + \text{ad inf} \qquad \qquad ...(2)$$

■ **57.** We can easily show that this series is convergent when x is complex.

For let $x = r(\cos\theta + \sqrt{-1}\sin\theta)$.

Then $$e^x = 1 + x + \frac{x^2}{\underline{|2}} + \frac{x^3}{\underline{|3}} + ... \text{ ad inf.}$$

$$= 1 + r(\cos\theta + i\sin\theta) + \frac{r^3(\cos 2\theta + i\sin 2\theta)}{\underline{|2}}$$

$$+ \frac{r^2(\cos 3\theta + i\sin 3\theta)}{\underline{|3}} + ... \text{ ad inf.}$$

$$= 1 + r\cos\theta + \frac{r^2\cos 2\theta}{\underline{|2}} + \frac{r^3\cos 3\theta}{\underline{|3}} + ...$$

$$+ \sqrt{-1}\left[r\sin\theta + \frac{r^2\sin 2\theta}{\underline{|2}} + \frac{r^3\sin 3\theta}{\underline{|3}} + ... \right].$$

The quantity

$$1 + r\cos\theta + \frac{r^2}{\underline{|2}}\cos 2\theta + \frac{r^3}{\underline{|3}}\cos 3\theta + ... \text{ ad inf.}$$

is $$< 1 + r + \frac{r^2}{\underline{|2}} + \frac{r^3}{\underline{|3}} + ... \text{ ad inf}$$

60

and is therefore convergent since this series is convergent for all real values of r. (Art 6.)

Similarly the quantity

$$r\sin\theta + \frac{r^2}{\underline{|2}}\sin2\theta + \ldots\ldots$$

is convergent.

Hence the series for e^x is always convergent.

▨ **58.** When x is a complex quantity the quantity e^x is then a *short way* of writing

$$1 + x + \frac{x^2}{\underline{|2}} + \frac{x^3}{\underline{|3}} + \ldots\ldots$$

Unless x be **real**, the e in e^x does not mean the series

$$1 + 1 + \frac{1}{\underline{|2}} + \frac{1}{\underline{|3}} + \ldots\ldots$$

When x is complex, e^x stands for a series of *the same form* as that series which, when x is real, has been proved to be equal to

$$\left(1 + 1 + \frac{1}{\underline{|2}} + \frac{1}{\underline{|3}} + \ldots\ldots\right)^x.$$

Instead of e^x the expressions $E(x)$ and $\exp(x)$ are sometimes used.

▨ **59.** By a proof similar to that of Art. 304, C. Smith's *Algebra*, it may be shown that

$$e^x \cdot e^y = e^{x+y},$$

whether x and y be real or complex quantities, so that the functions e^x and e^y obey a law of the same form as the index law.

▨ **60.** If x be put equal to θi, where θ is real, we then have

$$e^{\theta i} = 1 + \theta i + \frac{\theta^2 i^2}{\underline{|2}} + \frac{\theta^3 i^3}{\underline{|3}} + \ldots\ldots$$

$$= 1 - \frac{\theta^2}{\underline{|2}} + \frac{\theta^4}{\underline{|4}} - \frac{\theta^6}{\underline{|6}} + \ldots\ldots$$

$$+ i\left[\theta - \frac{\theta^3}{\underline{|3}} + \frac{\theta^5}{\underline{|5}} - \ldots\ldots\right]$$

$$= \cos\theta + i\sin\theta. \qquad\qquad \text{(Arts. 32 and 33.)}$$

So, $\qquad\qquad e^{-\theta i} = \cos\theta - i\sin\theta.$

Hence, by addition, we have

$$\cos\theta = \frac{e^{\theta i} + e^{-\theta i}}{2},$$

and, by subtraction,

$$\sin\theta = \frac{e^{\theta i} - e^{-\theta i}}{2i}.$$

Circular functions of complex angles.

■ **61.** When x is a complex quantity, the functions $\sin x$ and $\cos x$ have at present no meaning.

For **real** values of x we have already shown in Arts. 32 and 33 that

$$\sin x = x - \frac{x^3}{\underline{3}} + \frac{x^5}{\underline{5}} - \frac{x^7}{\underline{7}} + \dots \dots \text{ ad inf.}$$

and

$$\cos x = 1 - \frac{x^2}{\underline{2}} + \frac{x^4}{\underline{4}} - \frac{x^6}{\underline{6}} + \dots \dots \text{ ad inf.}$$

Let us **define** $\sin x$ and $\cos x$, when x is **complex,** so that these relations may always be true, *i.e.,* for all values of x let

$$\sin x = x - \frac{x^3}{\underline{3}} + \frac{x^5}{\underline{5}} - \frac{x^7}{\underline{7}} + \dots \dots$$

...(1)

and

$$\cos x = 1 - \frac{x^2}{\underline{2}} + \frac{x^4}{\underline{4}} + \frac{x^6}{\underline{6}} + \dots \dots$$

...(2)

When x is complex, the quantities $\sin x$ and $\cos x$ are then only *short ways* of writing the series on the right-hand sides of (1) and (2).

■ **62.** We have then, for all values of x, real or complex,

$$\cos x + i\sin x = 1 + xi - \frac{x^2}{\underline{2}} - \frac{x^3 i}{\underline{3}} + \frac{x^4}{\underline{4}} \dots \dots$$

$$= 1 + xi + \frac{(xi)^2}{\underline{2}} + \frac{(xi)^3}{\underline{3}} + \frac{(xi)^4}{\underline{4}} \dots \dots$$

$$= e^{xi}.$$ (Art. 56)

So

$$\cos x - i\sin x = e^{-xi}.$$

Hence for all values of x, real or complex, we have

$$\cos x = \frac{e^{xi} + e^{-xi}}{2}, \text{ and } \sin x = \frac{e^{xi} - e^{-xi}}{2i}.$$

These results are known as Euler's Exponential Values.

■ **63.** We can now show that the Addition and Subtraction Theorems hold for imaginary angles, *i.e.,* that, whether x be real or complex, then

$$\sin (x + y) = \sin x \cos y + \cos x \sin y,$$
$$\cos (x + y) = \cos x \cos y - \sin x \sin y,$$
$$\sin (x - y) = \sin x \cos y - \cos x \sin y,$$

and

$$\cos (x - y) = \cos x \cos y + \sin x \sin y.$$

Since

$$\cos x = \frac{e^{xi} + e^{-xi}}{2} \text{ and } \sin x = \frac{e^{xi} - e^{-xi}}{2i},$$

we have $\sin x \cos y + \cos x \sin y$

$$= \frac{e^{xi} - e^{-xi}}{2i} \frac{e^{yi} + e^{-yi}}{2} + \frac{e^{xi} + e^{-xi}}{2} \frac{e^{yi} - e^{-yi}}{2i}$$

$$= \frac{e^{xi} \cdot 2e^{yi} - e^{xi-} \cdot 2e^{-yi}}{4i} = \frac{e^{(x+y)} - e^{-(x+y)i}}{2i} \qquad \text{(Art. 59)}$$

$$= \sin (x + y).$$

Similarly the other results may be proved.

■ **64.** It follows that all formulae which have been proved for real angles and which are founded on the Addition and Subtraction Theorems are also true when we substitute for the real angle any complex quantity.

For example, since

$$\cos 3\theta = 4 \cos^3 \theta - 3 \cos \theta,$$

where θ is real; it follows that

$$\cos 3 (x + yi) = 4 \cos^3 (x + yi) - 3 \cos (x + yi).$$

Again, since, by De Moivre's Theorem, we know that

$$\cos n\theta + i \sin n\theta$$

is always one of the values of

$$(\cos \theta + i \sin \theta)^n,$$

when θ is real and n has any value, it follows that

$$\cos n (x + yi) + i \sin n (x + yi)$$

is always one of the values of

$$[\cos (x + yi) + i \sin (x + yi)]^n.$$

■ **65. Periods of complex circular functions.** In equations (1) and (2) of Art. 63 let x be complex and let $y = 2\pi$.

Then $\sin (x + 2\pi) = \sin x \cos 2\pi + \cos x \sin 2\pi$

$$= \sin x,$$

and $\cos (x + 2\pi) = \cos x \cos 2\pi - \sin x \sin 2\pi$

$$= \cos x.$$

Hence $\sin x$ and $\cos x$ both remain the same when x is increased by 2π. Similarly they will remain the same when x is increased by

$$4\pi, 6\pi, ..., 2n\pi.$$

Hence, when x is complex, the expressions $\sin x$ and $\cos x$ are periodic functions whose period is 2π.

This corresponds with the results we have already found for real angles.

(Art. 61, Part I)

EXAMPLES X

Assuming that $\cos x = \dfrac{e^{xi} + e^{-xi}}{2}$ and $\sin x = \dfrac{e^{xi} - e^{-xi}}{2i}$ prove that, for all values of x and y, real or complex,

1. $\cos^2 x + \sin^2 x = 1.$ 2. $\cos(-x) = \cos x.$

3. $\sin(-x) = -\sin x.$ 4. $\cos 2x = \cos^2 x - \sin^2 x = 1 - 2\sin^2 x.$

5. $\sin 3x = 3 \sin x - 4 \sin^3 x.$ 6. $\cos x - \cos y = 2\sin \dfrac{x+y}{2} \sin \dfrac{y-x}{2}.$

7. $\sin x - \sin y = 2\cos \dfrac{x+y}{2} \sin \dfrac{x-y}{2}.$

Prove that

8. $\{\sin(a + \theta) - e^{ai} \sin \theta\}^n = \sin^n a e^{-n\theta i}.$

9. $\sin(a + n\theta) - e^{ai} \sin n\theta = e^{-n\theta i} \sin a.$

10. $\{\sin(a - \theta) + e^{\pm ai} \sin \theta\}^n = \sin^{n-1} a\{\sin(a - n\theta) + e^{\pm ai} \sin n\theta\}.$

■ **66.** In the formulae of Art. 62 if x be a pure imaginary quantity and equal to yi, we have, since

$$i^2 = -1,$$

$$\cos yi = \frac{e^{y_i \cdot i} + e^{-y_i \cdot i}}{2} = \frac{e^{-y} - e^{y}}{2} = \frac{e^{y} + e^{-y}}{2},$$

and $$\sin yi = \frac{e^{y_i \cdot i} - e^{-y_i \cdot i}}{2i} = \frac{e^{-y} - e^{y}}{2i} = i \cdot \frac{e^{-y} - e^{y}}{(2-1)},$$

$$= i \frac{e^{y} - e^{-y}}{2}.$$

■ **67. Hyperbolic functions. Def.** The quantity

$$\frac{e^{y} - e^{-y}}{2},$$

whether y be real or complex, is called the hyperbolic sine of y and is written sinh y.
Similarly the quantity

$$\frac{e^{y} + e^{-y}}{2},$$

is called the hyperbolic cosine of y and is written

cosh y.

[It will be observed that the values of sinh y and cosh y are obtained from the exponential expressions for sin y and cos y by simply omitting the i's.]

The hyperbolic tangent, secant, cosecant, and cotangent are obtained from the hyperbolic sine and cosine just as the ordinary tangent, secant, cosecant and cotangent are obtained from the ordinary sine and cosine.

Thus, $$\tanh y = \frac{\sinh y}{\cosh y} = \frac{e^{y} - e^{-y}}{e^{y} + e^{-y}},$$

$$\operatorname{cosech} y = \frac{1}{\sinh y} = \frac{2}{e^{y} - e^{-y}},$$

$$\operatorname{sech} y = \frac{1}{\cosh y} = \frac{2}{e^{y} + e^{-y}},$$

and $$\coth y = \frac{1}{\tanh y} = \frac{e^{y} + e^{-y}}{e^{y} - e^{-y}}.$$

The hyperbolic cosine and sine have the same relation to the curve called the rectangular hyperbola that the ordinary circular cosine and sine have to the circle. Hence the use of the word hyperbolic.

■ **68.** From Arts. 66 and 67 we clearly have

$$\cos (yi) = \cosh y,$$

and $$\sin (yi) = i \sinh y,$$

So $$\tan (yi) = i \tanh y.$$

■ **69.** Corresponding to most general trigonometrical formulae involving the ratios of angles there are formulae involving the hyperbolic ratios.

For example, we have, for all values of the angle x,

$$\cos^2 x + \sin^2 x = 1,$$

so that $$\cos^2 (yi) + \sin^2 (yi) = 1,$$

and hence, by the last article,

$$\cosh^2 y - \sinh^2 y = 1.$$

[This may be deduced independently from the definition of the hyperbolic functions. For

$$\cosh^2 y - \sinh^2 y = \left(\frac{e^{y} + e^{-y}}{2} \right)^2 - \left(\frac{e^{y} - e^{-y}}{2} \right)^2$$

$$= \frac{e^{2y} + 2 + e^{-2y}}{4} - \frac{e^{3y} - 2 + e^{-3y}}{4} = 1.]$$

Again, for all values of u and v we have

$$\sin (u + v) = \sin u \cos v + \cos u \sin v.$$

Put $$u = xi \text{ and } v = yi,$$

so that

$$\sin [(x + y)i] = \sin (xi) \cos (yi) + \cos (xi) \sin (yi).$$

The expressions of the last article then give

$$i \sinh (x + y) = i \sinh x \cosh y + \cosh x \times i \sinh y,$$

∴ $$\sinh (x + y) = \sinh x \cosh y + \cosh x \sinh y.$$

[Directly from the definition of the hyperbolic ratios, we have

$$\sinh x \cosh y + \cosh x \sinh y$$

$$= \frac{e^x - e^{-x}}{2} \cdot \frac{e^y + e^{-y}}{2} + \frac{e^x + e^{-x}}{2} \cdot \frac{e^y - e^{-y}}{2} = \frac{2e^{x+y} - 2e^{-(x+y)}}{4},$$

on multiplication, $= \sinh (x + y).$]

Again, for all values of θ we have

$$\tan 3\theta = \frac{3 \tan \theta - \tan^3 \theta}{1 - 3 \tan^2 \theta}.$$

Put then $\theta = xi$, and we have

$$\tan (3xi) = \frac{3 \tan (xi) - \tan^3 (xi)}{1 - 3 \tan^2 (xi)}.$$

Hence the substitutions of Art. 314 give

$$i \tanh (3x) = \frac{3i \tanh x - i^3 \tanh^3 x}{1 - 3i^2 \tanh^2 x}$$

$$= \frac{3i \tanh x + i \tanh^3 x}{1 + 3 \tanh^2 x},$$

so that

$$\tanh (3x) = 3 \frac{\tanh x + \tanh^3 x}{1 + 3 \tanh^2 x}.$$

As before, this may be easily proved from the definition of tanh x.

■ **70.** In general it follows from (1) of Art. 68 that any general formula which is true for cosines of angles is also true if instead of cos we read cosh.

From (2) of the same article, since

$$\sin^2 (yi) = -\sinh^2 y,$$

it follows that any general formula involving the cosine and square of the sine of an angle is true if for cos we read cosh and for $\sin^2$ we read $- \sinh^2$.

Similarly, from (3) we may turn a formula involving $\tan^2$ into another by writing for $\tan^2$ the quantity $-\tanh^2$.

In this manner, formulae and series involving the hyperbolic functions may be obtained from 27, 28, 30, 44, 46, and 48–53 and also from Part I, Arts 241 and 242.

■ **71.** From the values in Art. 67 it follows, by Art. 56 that

$$\cosh x = \frac{1}{2}\left(e^x + e^{-x}\right)$$

$$= 1 + \frac{x^2}{\lfloor 2} + \frac{x^4}{\lfloor 4} + \frac{x^6}{\lfloor 6} + \ldots\ldots,$$

$$\sinh x = \frac{1}{2}\left[e^x - e^{-x}\right]$$

$$= x + \frac{x^3}{\underline{3}} + \frac{x^5}{\underline{5}} + \frac{x^7}{\underline{7}} + \ldots$$

These are the expansional values of cosh x and sinh x.

■ *72. *Periods of the hyperbolic functions.*

For all values of θ, real or complex, we have cos θi = cosh θ.

Hence

$$\cosh (x + yi) = \cos \{(x + yi)i\} = \cos (xi - y) = \cos [-2\pi + xi - y](\text{Art. } 65)$$
$$= \cos [(2\pi i + x + yi)i] = \cosh [2\pi i + x + yi]$$
$$= (\text{similarly}) \cosh [4\pi i + x + yi] = \ldots$$

Hence, the hyperbolic cosine is periodic, its period being imaginary and equal to $2\pi i$.

Again, since sinh $\theta = -i \sin \theta i$, we have

$$\sinh (x + yi) = -i \sin \{(x + yi)i\} = -i \sin [xi - y]$$
$$= -i \sin [-2\pi + xi - y] = -i \sin \{[2\pi i + x + yi]i\}$$
$$= \sinh [2\pi i + x + yi],$$

so that the period of sinh $(x + yi)$ is $2\pi i$.

Similarly it may be shown that the period of tanh $(x + yi)$ is πi.

The hyperbolic functions therefore differ from the circular functions in having no real period; their period is imaginary.

▌ 73. **EXAMPLE 1.** *Separate into its real and imaginary parts the expression* sin $(\alpha + \beta i)$.

We have $\qquad \sin (\alpha + \beta i) = \sin \alpha \cos \beta i + \cos \alpha \sin \beta i$

$$= \sin \alpha \frac{e^{\beta} + e^{-\beta}}{2} + \cos \alpha \frac{e^{-\beta} - e^{\beta}}{2i}$$

$$= \sin \alpha \frac{e^{\beta} + e^{-\beta}}{2} + i \cos \alpha \frac{e^{\beta} - e^{-\beta}}{2}$$

$$= \sin \alpha \cosh \beta + i \cos \alpha \sinh \beta.$$

▌ **EXAMPLE 2.** *Separate into its real and imaginary parts the expression* tan $(\alpha + \beta i)$.

We have $\qquad \tan (\alpha + \beta i) = \dfrac{\sin (\alpha + \beta i)}{\cos (\alpha + \beta i)}$

$$= \frac{2 \sin (\alpha + \beta i) \cos (\alpha - \beta i)}{2 \cos (\alpha + \beta i) \cos (\alpha - \beta i)}$$

$$= \frac{\sin 2\alpha + \sin 2\beta i}{\cos 2\alpha + \cos 2\beta i}$$

$$= \frac{\sin 2\alpha + i \sinh 2\beta}{\cos 2\alpha + \cosh 2\beta}. \qquad\qquad \text{(Art. 68.)}$$

Alter. Let tan $(\alpha + \beta i) = x + yi$, so that tan $(\alpha - \beta i) = x - yi$.

$$x = \frac{1}{2}\left[\tan(\alpha + \beta i) + \tan(\alpha - \beta i)\right]$$

$$= \frac{\sin(\alpha + \beta i)\cos(\alpha - \beta i) + \cos(\alpha + \beta i)\sin(\alpha - \beta i)}{2\cos(\alpha + \beta i) \cdot \cos(\alpha - \beta i)}$$

$$= \frac{\sin 2\alpha}{\cos 2\alpha + \cos 2\beta i} = \frac{\sinh 2\alpha}{\cos 2\alpha + \cosh 2\beta}$$

Also,
$$y = \frac{1}{2i}\left[\tan(\alpha + \beta i) - \tan(\alpha - \beta i)\right]$$

$$= \frac{1}{2i}\frac{\sin(\alpha + \beta i)\cos(\alpha - \beta i) - \cos(\alpha + \beta i)\sin(\alpha - \beta i)}{\cos(\alpha + \beta i)\cos(\alpha - \beta i)}$$

$$= \frac{1}{i}\frac{\sin 2\beta i}{\cos 2\alpha + \cos 2\beta i} = \frac{\sinh 2\beta}{\cos 2\alpha + \cosh 2}.$$

$$\therefore \qquad \tan(\alpha + \beta i) = \frac{\sin 2\alpha + i\sinh 2\beta}{\cos 2\alpha + \cosh 2\beta}.$$

▎**EXAMPLE 3.** *Separate into its real and imaginary parts the expression* cosh $(\alpha + \beta i)$.

We have
$$\cosh(\alpha + \beta i) = \frac{e^{\alpha + \beta i} + e^{-\alpha - \beta i}}{2} \qquad \text{(Art. 67)}$$

$$= \frac{e^{\alpha} \cdot e^{\beta i} + e^{-\alpha} \cdot e^{-\beta i}}{2} = \frac{e^{\alpha}(\cos\beta + i\sin\beta) + e^{-\alpha}(\cos\beta - i\sin\beta)}{2} \qquad \text{(Art. 62)}$$

$$= \frac{\cos\beta(e^{\alpha} + e^{-\alpha}) + i\sin\beta(e^{\alpha} - e^{-\alpha})}{2} = \cos\beta\cosh\alpha + i\sin\beta\sinh\alpha.$$

Alter. cosh $(\alpha + \beta i) = \{(\alpha + \beta i)i\}$ (Art. 68)

             $= \cos\{\alpha i - \beta\}$

             $= \cos(\alpha i)\cos\beta + \sin(\alpha i)\sin\beta$

             $= \cosh\alpha\cos\beta + i\sinh\alpha\sin\beta$.

EXAMPLES XI

Prove that

1. $\cosh 2x = 1 + 2(\sinh x)^2 = 2(\cosh x)^2 - 1$

2. $\cosh(\alpha + \beta) = \cosh\alpha\cosh\beta + \sinh\alpha\sinh\beta$.

3. $\cosh(\alpha + \beta i) - \cosh(\alpha - \beta) = 2\sinh\alpha\sinh\beta$

4. $\tanh(\alpha + \beta) = \dfrac{\tanh\alpha + \tanh\beta}{1 + \tanh\alpha\tanh\beta}$.

5. $\cosh 3x = 4\cosh^2 x - 3\cosh x$.

6. $\sinh 3x = 3\sinh x + 4\sinh^3 x$.

7. $\sinh (x + y) \cosh (x - y) = \frac{1}{2}(\sinh 2x + \sinh 2y)$.

8. $\cosh 2x + \cosh 5x + \cosh 8x + \cosh 11x$

$$= 4\cosh \frac{13x}{2} \cosh 3x \cosh \frac{3x}{2}.$$

9. $\cosh x + \cosh (x + y) + \cosh (x + 2y) + \dots$ to n terms

$$= \frac{\cosh\left(x + \dfrac{n-1}{2}y\right)\sinh \dfrac{ny}{2}}{\sinh \dfrac{y}{2}}.$$

10. $\sinh x + \sinh (x + y) + \sinh (x + 2y) + \dots$ to n terms

$$= \frac{\sinh\left(x + \dfrac{n-1}{2}y\right)\sinh \dfrac{ny}{2}}{\sin \dfrac{y}{2}}.$$

11. $\sinh x + n \sinh 2x + \dfrac{n(n-1)}{1 \cdot 2} \sinh 3x + \dots\dots$ to $(n + 1)$ terms

$$= 2^n \cosh^n \frac{x}{2} \sinh\left(\frac{n}{2}+1\right)x.$$

12. $\sinh \beta \sin \alpha + i \cosh \beta \cos \alpha = i \cos (\alpha + \beta i)$.

13. $\sin 2\alpha + i \sinh 2\beta = 2 \sin (\alpha + i\beta) \cos (\alpha - i\beta)$.

14. $\cos (\alpha + i\beta) + i \sin (\alpha + i\beta) = e^{-\beta} (\cos \alpha + i \sin \alpha)$.

15. If $\tan y = \tan \alpha \tanh \beta$, and $\tan z = \cot \alpha \tanh \beta$, then prove that $\tan (y + z) = \sinh 2\beta \,\operatorname{cosec} 2\alpha$.

16. If $u = \log \tan\left(\dfrac{\pi}{4} + \dfrac{\theta}{2}\right)$, prove that $\tanh \dfrac{u}{2} = \tan \dfrac{\theta}{2}$.

Separate into their real and imaginary parts the quantities

17. $\cos (\alpha + \beta i)$.   **18.** $\cot (\alpha + \beta i)$.

19. $\operatorname{cosec} (\alpha + \beta i)$.   **20.** $\sec (\alpha + \beta i)$.

21. $\sinh (\alpha + \beta i)$.   **22.** $\tanh (\alpha + \beta i)$.

23. $\operatorname{sech} (\alpha + \beta i)$.

24. Prove that $\tan \dfrac{u + iv}{2} = \dfrac{\sin u + i \sinh v}{\cos u + \cosh v}$.

25. If $\sin (A + iB) = x + iy$, prove that

$$\frac{x^2}{\cosh^2 B} + \frac{y^2}{\sinh^2 B} = 1, \text{ and } \frac{x^2}{\sin^2 A} + \frac{y^2}{\cos^2 A} = 1.$$

26. If $\tan (A + iB) = x + iy$, prove that

$$x^2 + y^2 + 2x \cot 2A = 1, \text{ and } x^2 + y^2 - 2y \coth 2B + 1 = 0.$$

27. If $\sin (\theta + \phi i) = \cos \alpha + i \sin \alpha$, prove that $\cos^2 \theta = \pm\sin \alpha$.

28. If $\sin (\theta + \phi i) = \rho(\cos \alpha + i \sin \alpha)$, prove that

$$\rho^2 = \frac{1}{2}[\cosh 2\phi - \cos 2\theta] \text{ and } \tan \alpha = \tanh \phi \cot \theta.$$

29. If $\cos(\theta + \phi i) = R(\cos \alpha + i \sin \alpha)$, prove that

$$\phi = \frac{1}{2} \log \frac{\sin(\theta - \alpha)}{\sin(\theta + \alpha)}.$$

30. If $\tan(\theta + \phi i) = \tan \alpha + i \sec \alpha$, prove that $e^{3\phi} = \pm \cot \dfrac{\alpha}{2}$, and that $2\theta = n\pi + \dfrac{\pi}{2} + \alpha$.

31. If $\tan(\theta + \phi i) = \cos \alpha + i \sin \alpha$, prove that

$$\theta = \frac{n\pi}{2} + \frac{\pi}{4}, \text{ and } \phi = \frac{1}{2} \log \tan\left(\frac{\pi}{4} + \frac{a}{2}\right).$$

32. If $A + iB = c \tan(x + iy)$, then

$$\tan 2x = \frac{2cA}{c^2 - A^2 - B^2}.$$

33. If $\tan(\theta + \phi i) = \sin(x + iy)$, then

$\coth y \sinh 2\phi = \cot x \sin 2\theta$.

34. If $\tan(\alpha + i\beta) = i$, α and β being real, prove that α is indeterminate and β is infinite.

Prove that

35. $\frac{1}{2}(\sinh x + \sin x) = x + \dfrac{x^5}{\lfloor 5} + \dfrac{x^9}{\lfloor 9} + \dots$ ad inf.

36. $\frac{1}{2}(\cosh x + \cos x) = 1 + \dfrac{x^4}{\lfloor 4} + \dfrac{x^8}{\lfloor 8} + \dots$ ad inf.

■ ****74. Inverse Circular Functions.** When α and β are real and $\alpha = \cos \beta$, we defined, in Art. 237, Part I., the inverse cosine of α to be that value of β which lies between 0 and π, and it was pointed out that β was a many-valued quantity.

If now $\qquad x + yi = \cos(u + vi)$,

then similarly $u + vi$ is said to be an inverse cosine of $x + yi$

But since

$$x + yi = \cos(u + vi) = \cos[2n\pi \pm (u + vi)] \qquad \text{(Art. 65)}$$

it follows that $2n\pi \pm (u + vi)$ is also an inverse cosine of $x + yi$, where n is any integer.

The inverse cosine of $x + yi$ is hence a many-valued function. When the many-valuedness of the inverse cosine is considered it is written

$$\cos^{-1}(x + yi).$$

The principal value of the inverse cosine of $x + yi$ is that value of $2n\pi \pm (u + vi)$ which is such that either $2n\pi + u$ or $2n\pi - u$ lies between 0 and π.

This principal value is denoted by $\cos^{-1}(x + yi)$.

We have then

$$\cos^{-1}(x + yi) = 2n\pi \pm \cos^{-1}(x + yi).$$

■ ****75.** Similarly if

$$x + yi = \sin(u + vi) = \sin\{n\pi + (-1)^n(u + vi)\},$$

then $n\pi + (-1)^n(u + vi)$ is an inverse sine of $x + yi$. It is a many-valued quantity and is denoted by $\sin^{-1}(x + yi)$. Its principal value is such that its real part lies between $-\dfrac{\pi}{2}$ and $\dfrac{\pi}{2}$, and is denoted by $\sin^{-1}(x + yi)$.

We then have

$$\sin^{-1}(x + yi) = n\pi + (-1)^n \sin^{-1}(x + yi).$$

Similarly $\tan^{-1}(x + yi)$ and $\tan^{-1}(x + yi)$ are defined, so that the principal value

of $\tan^{-1}(x + yi)$ is such that its real part lies between $-\dfrac{\pi}{2}$ and $+\dfrac{\pi}{2}$, and

$$\tan^{-1}(x + yi) = n\pi + \tan^{-1}(x + yi).$$

Similarly,

$$\sec^{-1}(x + yi) = 2n\pi \pm \sec^{-1}(x + yi),$$

$$\csc^{-1}(x + yi) = n\pi + (-1)^n \csc^{-1}(x + yi),$$

and
$$\cot^{-1}(x + yi) = n\pi + \cot^{-1}(x + yi).$$

▥ **76.** We shall henceforward use $\sin^{-1}$, $\sin^{-1}$, $\cos^{-1}$, $\cos^{-1}$, ...with the meanings above assigned.

▥ **77. Inverse hyperbolic functions.** If $x = \cosh y$ then similarly, as in Art. 74, we write $y = \cosh^{-1} x$.

If x be real, we have

$$x = \frac{e^y + e^{-y}}{2},$$

so that
$$e^{2y} - 2xe^y + 1 = 0,$$

and hence
$$e^y = x \pm \sqrt{x^2 - 1}$$

$$= x + \sqrt{x^2 - 1} \ \text{or} \ \frac{1}{x + \sqrt{x^2 - 1}},$$

∴
$$y = \pm \log\left(x + \sqrt{x^2 - 1}\right).$$

The positive value of the right-hand side is the one always taken.

Hence, when x is real, $\cosh^{-1} x$ is a single-valued function.

Similarly, $\sinh^{-1} x$ and $\tanh^{-1} x$ are defined; they are single-valued functions, when x is real.

▥ **78.** If $\alpha + \beta i = \cosh(x + yi)$, then $x + yi$ is said to be an inverse hyperbolic cosine of $\alpha + \beta i$.

But $\cosh(x + yi) = \cosh\{2n\pi i \pm (x + yi)\}$, as in Art. 72.

Hence $2n\pi i \pm (x + yi)$ is an inverse hyperbolic cosine of $\alpha + \beta i$. Its principal value is that value whose imaginary part lies between 0 and πi, $i\,e$, such that $2n\pi \pm y$ lies between 0 and π.

Similarly, the inverse hyperbolic sine and tangent of $\alpha + \beta i$ are defined. In this

case the principal values are such that the imaginary part lies between $-\dfrac{\pi}{2}i$ and $\dfrac{\pi}{2}i$.

▮ **79. EXAMPLE 1.** *Separate into real and imaginary parts the quantity*
$$\sin^{-1}(\cos \theta + i \sin \theta), \text{ where } \theta \text{ is real.}$$

Let $\qquad$ $\sin^{-1}(\cos\theta + i\sin\theta) = x + yi$,

so that $\qquad$ $\cos\theta + i\sin\theta = \sin(x + yi) = \sin x \cos yi + \cos x \sin yi$

$$= \sin x \cosh y + i\cos x \sinh y.$$

Hence, $\qquad$ $\sin x \cosh y = \cos\theta$ $\qquad$ ...(1)

and $\qquad$ $\cos x \sinh y = \sin\theta$ $\qquad$ ...(2)

Squaring and adding, we have

$$1 = \sin^2 x \cosh^2 y + \cos^2 x \sinh^2 y = \sin^2 x(1 + \sinh^2 y) + \cos^2 x \sinh^2 y = \sin^2 x + \sinh^2 y.$$

$\qquad\therefore\qquad$ $\sinh^2 y = \cos^2 x.$

Hence from (2) we have $\cos^2 x = \sin\theta$, assuming $\sin\theta$ to be positive.

Therefore, since x is to lie between $-\dfrac{\pi}{2}$ and $+\dfrac{\pi}{2}$ $\qquad$ (Art. 75)

we have $\cos x = +\sqrt{\sin\theta}$, and hence $x = \cos^{-1}(\sqrt{\sin\theta})$.

The equation (2) then gives

$$\sinh y = +\sqrt{\sin\theta},$$

so that $\qquad$ $e^{2y} - 2e^y \sqrt{\sin\theta} = 1$, a quadratic for e^y.

Hence $\qquad$ $e^y = \sqrt{\sin\theta} + \sqrt{1 + \sin\theta},$

i.e., $\qquad$ $y = \log[\sqrt{\sin\theta} + \sqrt{1 + \sin\theta}].$

▮ **Example 2:** *Separate into its real and imaginary parts the quantity*

$$\tan^{-1}(\alpha + \beta i).$$

Let $\qquad$ $\tan^{-1}(\alpha + \beta i) = (x + yi)$, so that $\tan(\alpha + yi) = \alpha + \beta i$,

and $\qquad$ $\tan(x - yi) = \alpha - \beta i.$

$\qquad\therefore\qquad$ $\tan 2x = \tan\{(x + yi) + (x - yi)\}$

$$= \frac{(\alpha + \beta i) - (\alpha - \beta i)}{1 - (\alpha + \beta i)(\alpha - \beta i)} = \frac{2\alpha}{1 - \alpha^2 - \beta^2},$$

$\qquad\therefore\qquad$ $x = \dfrac{1}{2}\tan^{-1}\dfrac{2\alpha}{1 - \alpha^2 - \beta^2}.$

Again $\quad \tan(2yi) = \tan[(x + yi) + (x - yi)]$

$$= \frac{(\alpha + \beta i) - (\alpha - \beta i)}{1 + (\alpha + \beta i)(\alpha - \beta i)} = \frac{2\beta i}{1 + \alpha^2 + \beta^2}.$$

$\qquad\therefore\qquad$ $i\dfrac{e^{2y} - e^{-2y}}{e^{2y} + e^{-2y}} = \dfrac{2\beta i}{1 + \alpha^2 + \beta^2}$ $\qquad$ ...(1)

$\qquad\therefore\qquad$ $\dfrac{e^{2y}}{e^{-2y}} = \dfrac{1 + \alpha^2 + \beta^2 + 2\beta}{1 + \alpha^2 + \beta^2 - 2\beta} = \dfrac{(1 + \beta)^2 + \alpha^2}{(1 - \beta)^2 + \alpha^2}.$

$$\therefore \qquad y = \frac{1}{4} \log \left\{ \frac{(1+\beta)^2 + \alpha^2}{(1-\beta)^2 + \alpha^2} \right\}.$$

Or again (1) gives $\tanh 2y = \dfrac{2\beta}{1+\alpha^2+\beta^2}$,

so that $\qquad y = \dfrac{1}{2} \tanh^{-1} \dfrac{2\beta}{1+\alpha^2+\beta^2}$.

We should have $\tan^{-1}(\alpha + \beta i) = n\pi + \tan^{-1}(\alpha + \beta i)$

$$= n\pi + \frac{1}{2} \tan^{-1} \frac{2\alpha}{1-\alpha^2-\beta^2} + \frac{i}{2} \tanh^{-1} \frac{2\beta}{1+\alpha^2+\beta^2}.$$

EXAMPLES XII

Separate into their real and imaginary parts the quantities

1. $\tan^{-1}(\cos\theta + i\sin\theta)$.

2. $\cos^{-1}(\cos\theta + i\sin\theta)$, where θ is a positive acute angle.

Prove that

3. $\sinh^{-1} x = \log\left(x + \sqrt{x^2+1}\right)$.

4. $\tanh^{-1} x = \sinh^{-1} \dfrac{x}{\sqrt{1-x^2}}$.

5. $\cosh^{-1} x = \log\left(\sqrt{x^2-1} + x\right)$.

6. $\tanh^{-1} x = \dfrac{1}{2} \log \dfrac{1+x}{1-x}$.

7. $\sin^{-1}(\operatorname{cosec}\theta) = \left\{2n + (-1)^n\right\}\dfrac{\pi}{2} + i(-1)^n \log\cot\dfrac{\theta}{2}$.

8. $\tan^{-1}\left(e^{\theta i}\right) = \dfrac{n\pi}{2} + \dfrac{\pi}{4} - \dfrac{i}{2} \log\tan\left(\dfrac{\pi}{4} - \dfrac{\theta}{2}\right)$.

9. $\tan^{-1} \dfrac{\tan 2\theta + \tanh 2\phi}{\tan 2\theta - \tanh 2\phi} + \tan^{-1} \dfrac{\tan\theta - \tanh\phi}{\tan\theta + \tanh\phi} = \tan^{-1}(\cot\theta\coth\phi)$.

When x is a real quantity, plot the graphs of

10. $\sinh x$ and $\operatorname{cosech} x$.

11. $\cosh x$ and $\operatorname{sech} x$.

12. $\tanh x$ and $\coth x$.

Logarithms of Complex Quantities

■ **80.** If $\alpha = e^x$, where α and x are real quantities, we know that x is called the logarithm of α to base e and we have shown in Art. 5 that

$$\alpha = e^x = 1 + x + \frac{x^2}{\lfloor 2} + \frac{x^3}{\lfloor 3} + \ldots\ldots \text{ ad inf.}$$

We may therefore look upon the logarithm, x, of α to base e as being derived as a root of the equation

$$\alpha = 1 + x + \frac{x^2}{\lfloor 2} + \frac{x^3}{\lfloor 3} + \ldots \text{ad inf.} \qquad \ldots (1)$$

As in other cases we shall now extend this result to complex quantities.

■ **81. Def.** *If $x + yi$ be any complex quantity and if $\alpha + \beta i$ be a quantity which is equal to e^{x+yi}, i.e. to the series*

$$1 + (x + yi) + \frac{(x + yi)^2}{\lfloor 2} + \frac{(x + yi)^3}{\lfloor 3} + \ldots\ldots,$$

then $x + yi$ is said to be a logarithm of $\alpha + \beta i$.

We say "a" logarithm because, as we shall now show, there are with the above definition many logarithms of a quantity.

We have $\qquad \alpha + \beta i = e^{x+yi} \qquad \ldots (1)$

Now, by Art. 62, we have, for all integral values of n.

$$e^{2n\pi i} = \cos 2n\pi + i\sin 2n\pi = 1 \qquad \ldots (2)$$

Hence, from (1) and (2) we have, by Art. 59,

$$\alpha + \beta i = e^{x+yi} \cdot e^{2n\pi i} = e^{x + (y + 2n\pi)i}.$$

According to the above definition we see that, if $x + yi$ be a logarithm of $\alpha + \beta i$, so also is

$$x + yi + 2n\pi i, \text{ i.e. } x + (y + 2n\pi)i.$$

■ **82.** We proceed to find the logarithms of the complex quantity $\alpha + \beta i$, where α and β are real.

By Art. 20, we have

$$\alpha + \beta i = r[\cos(2n\pi + \theta) + i \sin (2n\pi + \theta)],$$

74

where n is any integer, $r = +\sqrt{\alpha^2 + \beta^2}$, and θ is that value lying between $-\pi$ and $+\pi$ such that $\cos \theta$ is $\dfrac{\alpha}{r}$ and $\sin \theta$ is $\dfrac{\beta}{r}$, *i.e.* with the restriction of Art. 20,

$$\theta = \tan^{-1} \frac{\beta}{\alpha}.$$

If $x + yi$ be a logarithm of $\alpha + \beta i$, we have then

$$r[\cos(2n\pi + \theta) + i \sin (2n\pi + \theta)] = e^{x + yi}$$
$$= e^x \cdot e^{yi} \qquad \text{(Art. 59)}$$
$$= e^x (\cos y + i \sin y).$$

By equating real and imaginary parts, we have

$$e^x \cos y = r \cos (2n\pi + \theta),$$

and
$$e^x \sin y = r \sin (2n\pi + \theta).$$

Hence
$$e^x = r, \text{ and } y = 2n\pi + \theta.$$

Since x and r are both real, x is the ordinary algebraic Napierian logarithm of r, so that

$$x = \log_e r.$$

Hence, a logarithm of $\alpha + \beta i$ is

$$\log_e r + i (2n\pi + \theta),$$

i.e.
$$\log_e \sqrt{\alpha^2 + \beta^2} + i\left(2n\pi + \tan^{-1} \frac{\beta}{\alpha}\right).$$

Since n is any integer we see that there are therefore an infinite number of logarithms of $\alpha + \beta i$, and that these only differ by multiples of $2\pi i$.

■ **83.** With the extended definition of a logarithm given in Art. 81, it follows by the last article that the logarithm of any number is many-valued.

When this many-valued is taken into consideration we write the logarithm of $\alpha + \beta i$ as log $(\alpha + \beta i)$.

Hence,

$$\log(\alpha + \beta i) = \log_e \sqrt{\alpha^2 + \beta^2} + i\left(2n\pi + \tan^{-1} \frac{\beta}{\alpha}\right).$$

If we put n equal to zero in the value of log $(\alpha + \beta i)$ the result is called the principal value of the logarithm and is denoted by log $(\alpha + \beta i)$, so that

$$\log(\alpha + \beta i) - \log_e \sqrt{(\alpha^2 + \beta^2)} + i \tan^{-1} \frac{\beta}{\alpha},$$

and
$$\log (\alpha + \beta i) = 2n\pi i + \log(\alpha + \beta i).$$

The distinction between log and log is to be hereafter assumed.

■ **84.** *Any positive quantity has one real logarithm and an infinite number of imaginary ones.*

In the result of the preceding article put β equal to zero, and we have

$$\log \alpha = 2n\pi i + \log_e \alpha.$$

We therefore observe that, with our extended definition of a logarithm, every real quantity α has a real logarithm (which is equal to $\log_e \alpha$ as ordinarily defined) and an infinite number of imaginary logarithms, which are obtained by adding any multiple of $2\pi i$ to its real logarithm.

This might have been directly deduced from equation (1) of Art. 80. For this is an equation of infinite degree and therefore it has an infinite number of roots, of which only one is real.

It will be noted that the principal value of the logarithms (according to our extended definition) of a real number is equal to its ordinary algebraic logarithm.

■ **85.** *Logarithm of a negative quantity.* In the result of Art. 83 put $\beta = 0$, and $\alpha = -x$, where x is a real positive quantity.

$$\therefore \qquad +\sqrt{\alpha^2 + \beta^2} = +x, \text{ and } \tan^{-1}\frac{\beta}{\alpha}$$

[which is an angle such that its cosine is $\dfrac{-x}{+x}$ *i.e.* -1, and its sine zero (Art. 20)] is equal to π.

$$\therefore \qquad \log(-x) = 2n\pi i + \log_e x + \pi i,$$

and $\qquad \log(-x) = \log_e x + \pi i.$

Hence the principal value of the logarithm of a negative quantity $-x$ (with our extended definition) is equal to the ordinary algebraic logarithm of x added on to πi.

■ **86.** *Logarithm of a quantity which is wholly imaginary.*

In the result of Art. 83 put $\alpha = 0$, and we have

$$\log(\beta i) = 2n\pi i + \log_e \beta + i\frac{\pi}{2}$$

$$= \text{Log}_e \beta + i\left(2n + \frac{1}{2}\right)\pi,$$

so that the logarithm of any quantity which is wholly imaginary consists of two parts, the first of which is real, and the second of which is imaginary and many-valued.

As a particular case, put $\beta = 1$, and we have

$$\text{Log}(\sqrt{-1}) = i\left(2n + \frac{1}{2}\right)\pi,$$

so that the principal value of log $(\sqrt{-1})$ is $\dfrac{\pi}{2}i$.

■ **87.** In the result of Art. 83 put

$$\alpha = \cos\theta \text{ and } \beta = \sin\theta.$$

$$\therefore \qquad \log(\cos\theta + i\sin\theta)$$

$$= \log_e 1 + i(2n\pi + \theta) = \theta i + 2n\pi i,$$

$$\therefore \qquad \log e^{\theta i} = \theta i + 2n\pi i.$$

The principal value of log $e^{\theta i}$, *i.e.* log $e^{\theta i}$, is therefore that value of $(\theta + 2n\pi)i$ which is such that $\theta + 2n\pi$ lies between $-\pi$ and $+\pi$.

▨ **88. EXAMPLE 1.** *Resolve into its real and imaginary parts the expression*
$$\log \sin (x + yi).$$

Let $\log \sin (x + yi) = u + vi$, so that
$$e^{u + vi} = \sin (x + yi) = \sin x \cos yi + \cos x \sin yi$$

$$= \sin x \frac{e^{y} + e^{-y}}{2} + i \cos x \frac{e^{y} - e^{-y}}{2} \qquad \text{...(1)}$$

As in Art. 18 let the right-hand side of this expression equal
$$r[\cos(2n\pi + \theta) + i \sin (2n\pi + \theta)],$$

so that

$$r = +\sqrt{\sin^{2} x \left(\frac{e^{y} + e^{-y}}{2}\right)^{2} + \cos^{2} x \left(\frac{e^{y} - e^{-y}}{2}\right)^{2}}$$

$$= \frac{1}{2} \sqrt{(e^{2y} + e^{-2y}) - 2\cos 2x}$$

$$= \frac{1}{2} \sqrt{2\cosh 2y - 2\cos 2x} = \sqrt{\frac{\cosh 2y - \cos 2x}{2}},$$

and
$$\theta = \tan^{-1} \left[\cot x \frac{e^{y} - e^{-y}}{e^{y} + e^{-y}}\right] = \tan^{-1}[\cot x \tanh y],$$

with the usual restriction of Art. 20.

We have then from (1)
$$e^{u} (\cos v + i \sin v) = r[\cos(2n\pi + \theta) + i \sin (2n\pi + \theta)].$$

Hence $e^{u} = r$, so that $u = \log_{e} r$,

and
$$v = 2n\pi + \theta.$$

∴ $\quad \log \sin (x + yi) = u + vi = \log_{e} r + (2n\pi + \theta)i$

$$= \frac{1}{2} \log_{e} \left[\frac{\cosh 2y - \cos 2x}{2}\right] + i \left[2n\pi + \tan^{-1} (\cot x \tanh y)\right].$$

By putting n equal to zero, we have the principal value of
$$\log \sin (x + iy).$$

▮ **EXAMPLE 2.** *Find the general value of log* (-3).

Let $x + yi = \log (-3)$, so that
$$e^{x + yi} = -3.$$

Put
$$-3 = r\{\cos (2n\pi + \theta) + i \sin (2n\pi + \theta)\},$$

as in Art. 18.

Then we have $\quad r = 3$ and $\theta = \pi$.

Hence
$$3\{\cos (2n\pi + \pi) + i \sin (2n\pi + \pi)\}$$
$$= e^{x + yi} = e^{x} \cdot e^{yi} = e^{x} \{\cos y + i \sin y\}.$$

Hence $e^x = 3$, so that $x = \log_e 3$, and $y = 2n\pi + \pi$.

$\therefore$ $\qquad\qquad \log(-3) = \log_e 3 + (2n\pi + \pi)i$.

The principal value, obtained by putting n equal to zero, is

$$\log_e 3 + \pi i.$$

EXAMPLES XIII

Prove that

1. $\log(\cos\theta + i \sin\theta) = i\theta$, if $-\pi < \theta \not> \pi$.

2. $\log(-1) = \pi i$.

3. $\log(-i) = -\dfrac{\pi}{2}i$.

4. $\log(1 + \cos 2\theta + i \sin 2\theta) = \log_e(2\cos\theta) + i\theta$, if $-\pi < \theta \not> \pi$.

5. $\log\tan\left(\dfrac{\pi}{4} + \dfrac{\pi}{2}i\right) = i\tan^{-1}\sinh x$.

6. $\log\cos(x + yi) = \dfrac{1}{2}\log_e\left(\dfrac{\cosh 2y + \cos 2x}{2}\right) - i\tan^{-1}(\tan x\tanh y)$.

7. $\log\dfrac{\sin(x + yi)}{\sin(x - yi)} = 2i\tan^{-1}(\cot x\tanh y)$.

8. $\log\dfrac{\cos(x - yi)}{\cos(x + yi)} = 2i\tan^{-1}(\tan x\tanh y)$.

9. $i\log\dfrac{x - i}{x + i} = \pi - 2\tan^{-1}x$.

10. $\log(1 + i\tan\alpha) = \log_e\sec\alpha + \alpha i$, where α is a positive acute angle.

11. $\log\left(\dfrac{1}{1 - e^{\theta i}}\right) = \log_e\left(\dfrac{1}{2}\operatorname{cosec}\dfrac{\theta}{2}\right) + i\left(\dfrac{\pi}{2} - \dfrac{\theta}{2}\right)$.

12. $\log\dfrac{a + bi}{a - bi} = 2i\tan^{-1}\dfrac{b}{a}$.

13. $\log(-5) = \log_e 5 + (2n\pi + \pi)i$.

14. $\mathrm{Log}(1 + i) = \dfrac{1}{2}\log_e 2 + i\left(2n\pi + \dfrac{\pi}{4}\right)$.

15. Find the value of $\log\log\sin(x + yi)$.

■ **89. Definition of a^x when a and x are any quantities, complex or real.** When a and x are real quantities we know that

$$a^x = e^x \log_e a. \qquad\qquad \text{(Art. 5)}$$

When a and x are complex the ordinary algebraic definition of a^x no longer holds. Let us so define it that

$$a^x = e^{x\log a},$$

for all values of x and a, whether real or complex.

Now, by Art. 83, log a is many-valued and complex when a is complex. Hence a^x is many-valued and complex, so that

$$a^x = e^{x \log a} = e^{x (2n\pi i + \log a)}.$$

The value of a^x obtained by putting n equal to zero is called its principal value. Hence the principal value of a^x

$$= e^{x \log a}$$

$$= 1 + x \log a + \frac{x^2}{\lfloor 2} (\log a)^2 + \dots \qquad \dots\text{(by Art. 56)}.$$

From Art. 59 it follows that if principal values be considered we have $a^x \times a^y = a^{x+y}$, so that the principal value of a^x satisfies the ordinary algebraic law of indices.

■ **90.** It may now be shown that, if y be complex,

$$\log(1+y) = y - \frac{1}{2} y^2 + \frac{1}{3} y^3 - \frac{1}{4} y^4 + \dots \text{ ad inf.}$$

The proof is similar to the proof when y is real. (Art. 8)

It is, in general, necessary that the modulus of y be < 1; otherwise the Binomial Theorem does not hold for complex quantities. (Art. 26.)

If the modulus of y be equal to unity, so that y may be put equal to $\cos \phi + i \sin \phi$, the expansion can be shown to be still true, except in the cases when ϕ is equal to an odd multiple of π.

Since $\qquad \log (1 + y) = 2n\pi i + \log (1 + y),$

we have

$$\log(1+y) = 2n\pi i + y - \frac{1}{2} y^2 + \frac{1}{3} y^3 - \frac{1}{4} y^4 + \dots \text{ ad inf.}$$

■ **91.** *To separate into its real and imaginary parts the expression* $(\alpha + \beta i)^{x + yi}$.

Let $\qquad \alpha + \beta i = r (\cos \theta + i \sin \theta),$

so that, as in Art. 18,

$$r = \sqrt{\alpha^2 + \beta^2}, \text{ and } \theta = \tan^{-1} \frac{\beta}{\alpha}.$$

Then, by definition,

$$(\alpha + \beta i)^{x + yi} = e^{(x + yi) \log (\alpha + \beta i)}$$

$$= e^{(x + yi) \{\log (\alpha + \beta i) + 2m\pi i\}}$$

$$= e^{(x + yi) \{\log r + (\theta + 2m\pi)i\}}$$

$$= e^{\{x \log r - y (\theta + 2m\pi)\} + i\{y \log r + x (\theta + 2m\pi)\}}$$

$$= e^{x \log r} \cdot e^{- y(\theta + 2m\pi)} \cdot e^{i\{y \log r + x (\theta + 2m\pi)\}}$$

$$= r^x \cdot e^{- y(\theta + 2m\pi)} [\cos \{y \log r + x (\theta + 2m\pi)\}$$
$$+ i \sin \{y \log r + x (\theta + 2m\pi)\}].$$

If we put m equal to zero, we obtain the principal value of the given quantity, viz.

$$r^x e^{-y\theta}[\cos (y \log r + x\theta) + i \sin (y \log r + x\theta)].$$

■ ****92. EXAMPLE 1.** *Find the general value of* $[\sqrt{-1}]^{\sqrt{-1}}$.

We have $\qquad [\sqrt{-1}]^{\sqrt{-1}} = e^{\sqrt{-1}\,\text{Log}\sqrt{-1}}$.

But $\qquad \text{Log}\sqrt{-1} = \text{Log}\left[\cos\left(2n\pi + \dfrac{\pi}{2}\right) + i\sin\left(2n\pi + \dfrac{\pi}{2}\right)\right]$

$$= \text{Log}_e\left(2n\pi + \dfrac{\pi}{2}\right)^i = \left(2n\pi + \dfrac{\pi}{2}\right)i.$$

$\therefore \qquad [\sqrt{-1}]^{\sqrt{-1}} = e\left(2n\pi + \dfrac{\pi}{2}\right)^{i^2} = e^{-}\left(2n\pi + \dfrac{\pi}{2}\right),$

where n has any integral value.

The principal value of $[\sqrt{-1}]^{\sqrt{-1}}$ is $e^{-\frac{\pi}{2}}$.

■ **EXAMPLE 2.** *Find the general value of* $\log_2(-3)$.

Let $\qquad \log_2(-3) = x + yi$, so that $2^{x+yi} = -3$,

i.e. $e^{(x+yi)\log 2} = 3\{\cos(2m\pi + \pi) + i\sin(2m\pi + \pi)\}$ $\qquad$ (Art. 20).

But $\qquad \log 2 = 2n\pi i + \log_e 2$, and $3 = e^{\log_e^2}$,

$\therefore \qquad e^{(x+yi)(3n\pi i + \log_e 2)} = e^{\log_2 3} \cdot e^{(2m\pi + \pi)i}$.

$\therefore \qquad (x + yi)(2n\pi i + \log_e 2) = \log_e 3 + (2m\pi + \pi)i.$

Equating real and imaginary parts, we have

$$x\log_e 2 - 2n\pi y = \log_e 3,$$

and $\qquad x \cdot 2n\pi + y\log_e 2 = 2m\pi + \pi.$

Solving, we have

$$x = \frac{\log_e 3\log_e 2 + (2m\pi + \pi)\cdot 2n\pi}{(\log_e 2)^2 + 4n^2\pi^2}$$

and $\qquad y = \dfrac{(2m\pi + \pi)\log_e 2 - 2n\pi\log_e 3}{(\log_e 2)^2 + 4n^2\pi^2},$

Hence, $\qquad \text{Log}_2(-3)$

$$= \frac{\left\{\log_e 3\log_e 2 + 2n(2m+1)n^2\right\} + i\pi\left\{(2m+1)\log_e 2 - 2n\log_e 3\right\}}{(\log_e 2)2^2 + 4n^2\pi^2}$$

If $m = n = 0$, the principal value is obtained, *viz.*

$$\frac{\log_e 3 + \pi i}{\log_e 2}.$$

■ **93.** It could now be shown that the general values of the logarithms of complex quantities satisfy the ordinary laws of logarithms, *viz.*

$$\log mn = \log m + \log n,$$

and
$$\log \frac{m}{n} = \log m - \log n.$$

It could also be shown that $\log m^n = n \log m + 2\,p\pi i$, where p is some integer or zero. The proof is left as an exercise for the student.

EXAMPLES XIV

Prove that

1. $a^i = e^{-2m\pi}\{\cos (\log a) + i \sin (\log a)\}.$

2. $i^a = \cos\left\{\left(2m + \frac{1}{2}\right)\pi a\right\} + i\sin\left\{\left(2m + \frac{1}{2}\right)\pi a\right\}.$

3. $i^{i^{-1}} = \cos\theta + i\sin\theta,$ where
$$\theta = \left(2m + \frac{1}{2}\right)\pi \cdot e^{-\left(2m\pi + \frac{\pi}{2}\right)}.$$

4. If $i^{i^{-1}}$ ...ad inf $= A + Bi$, principal values only being considered, prove that
$$\tan \frac{\pi A}{2} = \frac{B}{A}, \text{ and } A^2 + B^2 = e^{-\pi B}.$$

5. If $i^{\alpha+\beta i} = \alpha + \beta i$, prove that
$$\alpha^2 + \beta^2 = e^{-(4n+1)\pi\beta}.$$

6. $\dfrac{(1+i)^{p+q_i}}{(1-i)^{p-q_i}} = \alpha + \beta i,$ prove that one value of $\tan^{-1}\dfrac{\beta}{\alpha}$ is
$$\frac{1}{2}p\pi + q\log_e 2.$$

7. If $(a + bi)^p = m^{x+yi}$, prove that one of the values of $\dfrac{y}{x}$ is
$$\frac{2\tan^{-1}\dfrac{b}{u}}{\log_e\left(a^2 + b^2\right)}.$$

8. If $a^{\alpha+\beta i} = (x + yi)^{p+qi}$, principal values only being considered, prove that
$$a = \frac{1}{2}p\log_u\left(x^2 + y^2\right) - q\tan^{-1}\frac{y}{x}\log_u e_y$$

and that
$$\log_a\left(x^2 + y^2\right) = 2\frac{\alpha p + \beta q}{p^2 + q^2}.$$

9. Prove that the real part of the principal value of (i) $\log(1+i)$ is
$$e^{-\frac{\pi^2}{8}}\cos\left(\frac{\pi}{4}\log 2\right).$$

10. Prove that the principal value of $(a + ib)^{\alpha+i\beta}$ is wholly real or wholly imaginary according as

$$\tfrac{1}{2}\beta\log(a^2 + b^2) + \alpha\tan^{-1}\frac{b}{a}$$

is an even or an odd multiple of $\dfrac{\pi}{2}$.

11. Prove that the general value of

$$(1 + i\tan\alpha)^{-1}$$

is $e^{n+2m\pi}\left[\cos\{\log\cos\alpha\} + i\sin\{\log\cos\alpha\}\right].$

12. If

$$\left(\frac{a+x+iy}{a-x-iy}\right)^{\lambda+\mu i} = X + i\gamma,$$

prove that one of the values of

$$\tan^{-1}\frac{\gamma}{X} \text{ is } \lambda\tan^{-1}\left(\frac{2ay}{a^2 - x^2 - y^2}\right) + \frac{\mu}{2}\log\frac{(a+x)^2 + y^2}{(a-x)^2 + y^2}.$$

13. Prove that $\log\sqrt{-1}^{(\sqrt{-1})} = \dfrac{4n+1}{4m+1}$,

where m and n are any integers.

14. Prove that the general value of $\log_4(-2)$ is

$$\frac{\log(2)^2 + m\cdot(2n+1)\pi^2}{2(\log 2)^2 + 2m^2\pi^2} + i\frac{(2n+1-m)\pi\log 2}{2(\log 2)^2 + 2m^2\pi^2}.$$

Explain the fallacies in the following arguments:

15. For all integral values of n we have

$$e^{2n+i} = \cos 2n\pi + i\sin 2n\pi = 1,$$

so that $\qquad e^{2ni} = e^{4ni} = e^{6ni} = ...$

Raise all these quantities to the power $\sqrt{-1}$; thus

$$e^{-2\pi} = e^{-4\pi} = e^{-6\pi} = ...$$

$$\therefore\ 2\pi = 4\pi = 6\pi = ...$$

16. For all values of θ we have

$$\cos(\theta - \pi) + i\sin(\theta - \pi) = \cos(\theta + \pi) + i\sin(\theta + \pi),$$

so that $\qquad e^{i(\theta - \pi)} = e^{i(\theta + \pi)}$

Hence, $\theta - \pi = \theta + \pi$, *i.e.* $\pi = 0$.

17. If θ and ϕ be the principal values of the amplitudes of two complex numbers x and y, prove that

$$\log xy = \log x + \log y + 2n\pi i,$$

where n is -1, 0, or $+1$ according as $\theta + \phi$ is $> \pi$ greater than $-\pi$ and not greater than π, and not greater than $-\pi$, respectively.

Gregory's Series
Calculation of the Value of π

■ **94. Gregory's Series.** *To prove that, if* θ *be not less than* $-\dfrac{\pi}{4}$ *and be not greater than* $+\dfrac{\pi}{4}$, *then*

$$\theta = \tan\theta - \frac{1}{3}\tan^3\theta + \frac{1}{5}\tan^5\theta - \dots \text{ ad inf.}$$

We have

$$1 + i\tan\theta = \sec\theta\,(\cos\theta + i\sin\theta)$$
$$= \sec\theta.\,e^{\theta i}.$$

Hence, by Art. 83, we have

$$\log_e \sec\theta + \theta i = \log(1 + i\tan\theta).$$

Therefore, by Art. 90, if tan θ be numerically not greater than unity, we have

$$\log_e(\sec\theta) + \theta i = \log(1 + i\tan\theta)$$

$$= i\tan\theta - \frac{1}{2}i^2\tan^2\theta + \frac{1}{3}i^3\tan^3\theta - \dots .$$

$$= i\tan\theta + \frac{1}{2}\tan^2\theta - \frac{1}{3}i\tan^3\theta - \frac{1}{4}\tan^4\theta + \dots \text{ ad inf.}$$

Equating the imaginary parts on each side of this equation, we have

$$\theta = \tan\theta - \frac{1}{3}\tan^3\theta + \frac{1}{5}\tan^5\theta - \frac{1}{7}\tan^7\theta + \dots \text{ ad inf.} \qquad \dots(1)$$

Since this series is true for acute angles such that the tangent is not numerically greater than unity it is true for all angles lying between the values $-\dfrac{\pi}{4}$ and $+\dfrac{\pi}{4}$ and also for the extreme values $-\dfrac{\pi}{4}$ and $+\dfrac{\pi}{4}$.

■ **95.** The series of the last article may be slightly transformed by writing tan θ = x, so that x must be not less than −1 and not greater than 1.

It then becomes

$$\tan^{-1}x = x - \frac{1}{3}x^3 + \frac{1}{5}x^5 - \frac{1}{7}x^7 + \dots \text{ad inf.},$$

83

where $\tan^{-1} x$ is that value which lies between

$$-\frac{\pi}{4} \text{ and } +\frac{\pi}{4}.$$

■ **96.** Gregory's Series is a particular case of a more general theorem which may be enunciated as follows:

If θ be an angle which lies between $p\pi - \frac{\pi}{4}$ and $p\pi + \frac{\pi}{4}$, both limits being admissible, then

$$\theta - p\pi = \tan\theta - \frac{1}{3}\tan^3\theta + \frac{1}{5}\tan^5\theta - ... \text{ ad inf.}$$

For let $\theta = p\pi + \phi$, where ϕ is not greater than $\frac{\pi}{4}$ and not less than $-\frac{\pi}{4}$.

Then $1 + i \tan\theta = 1 + i \tan\phi = \sec\phi (\cos\phi + i \sin\phi)$

$$= \sec\phi . e^{\phi i}.$$

Hence, by Arts. 83 and 90, we have, provided that $\tan\theta$ be numerically not greater than unity,

$$\log_e \sec\phi + \phi i = \log(1 + i \tan\theta)$$

$$= i\tan\theta - \frac{1}{2}i^2\tan^2\theta + \frac{1}{3}i^3\tan^3\theta - ...$$

$$= i\tan\theta + \frac{1}{2}\tan^2\theta - \frac{1}{3}i\tan^3\theta + \frac{1}{4}\tan^4\theta + \frac{1}{5}i\tan^5\theta - ... \text{ ad inf.}$$

Equating the imaginary parts on both sides of this equation we have

$$\phi = \tan\theta - \frac{1}{3}\tan^3\theta + \frac{1}{5}\tan^5\theta -,... \text{ ad inf.,}$$

i.e. $\qquad \theta - p\pi = \tan\theta - \frac{1}{3}\tan^3\theta + \frac{1}{5}\tan^5\theta - ... \text{ ad inf.} \qquad\qquad ... (1)$

■ **97.** *Examples of particular cases.*

If θ lie between $\frac{3\pi}{4}$ and $\frac{5\pi}{4}$, *i.e.* between $\pi - \frac{\pi}{4}$ and $\pi + \frac{\pi}{4}$, we have $p = 1$ and equation (1) of the preceding article becomes

$$\theta - \pi = \tan\theta - \frac{1}{3}\tan^3\theta + \frac{1}{5}\tan^5\theta - ... \text{ ad inf.}$$

If θ lie between $\frac{7\pi}{4}$ and $\frac{9\pi}{4}$, *i.e.* between $2\pi - \frac{\pi}{4}$ and $2\pi + \frac{\pi}{4}$, the equation becomes

$$\theta - 2\pi = \tan\theta - \frac{1}{3}\tan^3\theta + \frac{1}{5}\tan^5\theta - ... \text{ ad inf.}$$

Similarly, if θ lie between $-\frac{13\pi}{4}$ and $-\frac{11\pi}{4}$, *i.e.* between $-3\pi - \frac{\pi}{4}$ and $-3\pi + \frac{\pi}{4}$, we have $p = -3$, and the equation becomes

$$\theta + 3\pi = \tan\theta - \frac{1}{3}\tan^3\theta + \frac{1}{5}\tan^5\theta - \dots \text{ ad inf.}$$

■ **98.** If θ lie between $\dfrac{\pi}{4}$ and $\dfrac{3\pi}{4}$, or between

$$\frac{5\pi}{4} \text{ and } \frac{7\pi}{4}, \dots$$

or, generally, between

$$n\pi + \frac{\pi}{4} \text{ and } n\pi + \frac{3\pi}{4},$$

$\tan\theta$ is numerically greater than unity; in these cases the expansion of $\log(1 + i\tan\theta)$ does not hold, and there is no such expansion as equation (1) of Art. 96.

■ **99. Value of π.** One of the chief uses of Gregory's series is its application to find the value of π.

In Art. 95 put $x = 1$, and we have

$$\frac{\pi}{4} = 1 - \frac{1}{3} + \frac{1}{5} - \frac{1}{7} + \frac{1}{9} - \dots$$

$$= 1 - \left(\frac{1}{3} - \frac{1}{5}\right) - \left(\frac{1}{7} - \frac{1}{9}\right) - \left(\frac{1}{11} - \frac{1}{13}\right)\dots$$

$$= 1 - 2\left[\frac{1}{3.5} + \frac{1}{7.9} + \frac{1}{11.13} + \dots\right].$$

This series may be used to calculate π; its defect however is that the successive terms do not rapidly become small, so that a very large number of terms would have to be taken to obtain the value of π correct to any great degree of accuracy.

For this reason other series have been sought for.

■ **100. Euler's Series.** We can easily prove that

$$\tan^{-1}\frac{1}{2} + \tan^{-1}\frac{1}{3} = \frac{\pi}{4}.$$

In Art. 95 put in succession x equal to

$$\frac{1}{2} \text{ and } \frac{1}{3},$$

and we have

$$\frac{\pi}{4} = \tan^{-1}\frac{1}{2} + \tan^{-1}\frac{1}{3}$$

$$= \frac{1}{2} - \frac{1}{3}\cdot\frac{1}{2^3} + \frac{1}{5}\cdot\frac{1}{2^5} - \frac{1}{7}\cdot\frac{1}{2^7} + \dots$$

$$+ \frac{1}{3} - \frac{1}{3}\cdot\frac{1}{3^3} + \frac{1}{5}\cdot\frac{1}{3^5} - \frac{1}{7}\cdot\frac{1}{3^7} + \dots$$

This series converges more quickly than the preceding series; but more than eleven terms of the series for $\tan^{-1}\dfrac{1}{2}$ would have to be taken to give π correct to 7 places of decimals.

■ **101. Machin's Series.** A more convergent series than the preceding is Machin's, which is derived from the expression

$$4 \tan^{-1} \frac{1}{5} - \tan^{-1} \frac{1}{239} = \frac{\pi}{4} \qquad \text{(Art. 240, Part I., Ex. 4)}$$

By substituting in succession $\frac{1}{5}$ and $\frac{1}{239}$ for x in Art. 95 we have

$$\frac{\pi}{4} = 4\left[\frac{1}{5} - \frac{1}{3}\cdot\frac{1}{5^3} + \frac{1}{5}\cdot\frac{1}{5^5} - \frac{1}{7}\cdot\frac{1}{5^7} + \ldots \right]$$

$$- \left[\frac{1}{239} - \frac{1}{3}\cdot\frac{1}{239^3} + \frac{1}{5}\cdot\frac{1}{239^5} - \ldots \right].$$

∴
$$\pi = 16\left[\frac{2}{10} - \frac{1}{3}\cdot\frac{2^3}{10^3} + \frac{1}{5}\cdot\frac{2^5}{10^5} - \frac{1}{7}\cdot\frac{2^7}{10^7} + \ldots \right]$$

$$-4\left[\frac{1}{239} - \frac{1}{3}\cdot\frac{1}{239^3} + \frac{1}{5}\cdot\frac{1}{239^5} - \ldots \right].$$

Now
$$16 \times \frac{2}{10} = 3.2$$

$$16 \times \frac{1}{5}\cdot\frac{2^5}{10^5} = 0.001024$$

$$16 \times \frac{1}{9}\cdot\frac{2^9}{10^9} = 0.0000009102$$

$$\cdots\cdots\cdots\cdots\cdots\cdots\cdots\cdots\cdots\cdots\cdots\cdots$$

$$4 \times \frac{1}{3}\cdot\frac{1}{239^3} = 0.0000000977$$

$$\overline{\hspace{2cm} 3.2010250079}$$

Also
$$16 \times \frac{1}{3}\cdot\frac{2^3}{10^3} = 0.426666666\ldots$$

$$16 \times \frac{1}{7}\cdot\frac{2^7}{10^7} = 0.0000292571\ldots$$

$$16 \times \frac{1}{11}\cdot\frac{2^{11}}{10^{11}} = 0.0000000298\ldots$$

$$\cdots\cdots\cdots\cdots\cdots\cdots\cdots\cdots$$

$$4 \times \frac{1}{239} = 0.0167364017\ldots$$

$$\overline{\hspace{2cm} 0.0594323552}$$

Hence

$$3.2010250079$$
$$- 0.0594323552$$

$$\pi = 3.14159265/27$$

This is the value of π correct to 8 places of decimals.

By taking the first series to 21 terms and the second series to three terms we should get π correct to sixteen places.

▨ **102. Rutherford's Series.** A further simplification of Machin's formula is the expression

$$4\tan^{-1}\frac{1}{5} - \tan^{-1}\frac{1}{70} + \tan^{-1}\frac{1}{99} = \frac{\pi}{4}.$$

For we have

$$\tan^{-1}\frac{1}{70} - \tan^{-1}\frac{1}{99} = \tan^{-1}\frac{\dfrac{1}{70} - \dfrac{1}{99}}{1 + \dfrac{1}{70}\cdot\dfrac{1}{99}} = \tan^{-1}\frac{29}{6931}$$

$$= \tan^{-1}\frac{1}{239}.$$

EXAMPLES XV

Assuming that

$$\theta - n\pi = \tan\theta - \frac{1}{3}\tan^3\theta + \frac{1}{5}\tan^5\theta - ...,$$

Write down the value of n when θ lies between

1. $\dfrac{11\pi}{4}$ and $\dfrac{13\pi}{4}$

2. $\dfrac{7\pi}{4}$ and $\dfrac{9\pi}{4}$

3. $\dfrac{19\pi}{4}$ and $\dfrac{21\pi}{4}$

4. $-\dfrac{3\pi}{4}$ and $-\dfrac{5\pi}{4}$

5. $-\dfrac{11\pi}{4}$ and $-\dfrac{13\pi}{4}$

6. Prove that

$$\pi = 2\sqrt{3}\left\{1 - \frac{1}{3^2} + \frac{1}{5.3^2} - \frac{1}{7.3^3} + ...\right\}.$$

7. Prove that

$$\frac{\pi}{4} = \frac{2}{3} + \frac{1}{7} - \frac{1}{3}\left(\frac{2}{3^3} + \frac{1}{7^3}\right) + \frac{1}{5}\left(\frac{2}{3^5} + \frac{1}{7^5}\right) - ...$$

8. If x be $< \sqrt{2} - 1$, prove that

$$2\left(x - \frac{1}{3}x^3 + \frac{1}{5}x^5 \dots\dots \text{ad inf.}\right)$$

$$= \frac{2x}{1-x^2} - \frac{1}{3}\left(\frac{2x}{1-x^2}\right)^3 + \frac{1}{5}\left(\frac{2x}{1-x^2}\right)^3 - \dots \text{ad inf.}$$

Find the value of π to three places of decimals.

9. By using Euler's Series.

10. By using Machin's Series.

11. By using Rutherford's Series.

12. To the second order of small quantities, prove that

$$\frac{1}{2}\sqrt{1+\sin\theta}\,\log(1-\theta) + \tan^{-1}\theta\sin\left(\frac{\pi}{3}+\theta\right) = \frac{\sqrt{3}-1}{2}\theta.$$

13. When both θ and $\tan^{-1}(\sec\theta)$ lie between 0 and $\dfrac{\pi}{2}$, prove that

$$\tan^{-1}(\sec\theta) = \frac{\pi}{4} + \tan^2\frac{\theta}{2} - \frac{1}{3}\tan^6\frac{\theta}{2} + \frac{1}{5}\tan^{10}\frac{\theta}{2} - \dots\dots$$

CHAPTER 8

Summation of Series
Expansions in Series

■ **103.** We shall now apply the results of the preceding chapters to the summation of some trigonometrical series.

The chief series may be divided into four classes;

1. Those depending for their summation on a Geometrical Progression ultimately,
2. Those depending ultimately on the Binomial Theorem,
3. Those depending ultimately on the Exponential Theorem, including as sub-cases, the Sine and Cosine series, and
4. Those depending ultimately on the Logarithmic Series and, as a sub-case, Gregory's Series.

■ **104.** In Arts. 105–108 we shall sum one example of each of these classes. It will generally be found more convenient in summing one of these series involving *sines* of multiple angles (such as sin α, sin 2α, sin 3α...) to also sum at the same time the companion series involving the *cosines* of the same multiple angles.

(*i.e.*, cos α, cos 2α, cos 3α...).

The method will be best seen by a careful study of the following four articles.

▌ **105. EXAMPLE:** *Sum to n terms, and to infinity, the series*

$$1 + c \cos \alpha + c^2 \cos 2\alpha + ...,$$

where c is less than unity.

Let

$$C = 1 + c \cos \alpha + c^2 \cos 2\alpha + \quad + c^{n-1} \cos (n-1) \alpha \qquad ...(1)$$

and

$$S = c \sin \alpha + c^2 \sin 2\alpha + ... + c^{n-1} \sin (n-1) \alpha \qquad ...(2)$$

Multiplying (2) by i and adding to (1), we have

$$C + Si = 1 + c(\cos \alpha + i \sin \alpha) + c^2(\cos 2\alpha + i \sin 2\alpha) + ...$$

$$= 1 + ce^{ai} + c^2 e^{2ai} + ... + c^{n-1} e^{(n-1)ai} \qquad \text{(Art. 62)}$$

$$= \frac{1 - c^n e^{nai}}{1 - ce^{ai}}, \text{ by summing the G.P.,}$$

$$= \frac{(1 - c^n e^{nai})(1 - ce^{-ai})}{(1 - ce^{ai})(1 - ce^{-ai})}$$

$$= \frac{1 - ce^{-ai} - c^n e^{nai} + c^{n+1} e^{(n-1)ai}}{1 - c(e^{ai} + e^{-ai}) + c^2}$$

$$= \frac{1 - c(\cos\alpha + i\sin\alpha) - c^n(\cos n\alpha + i\sin n\alpha) + e^{n+1}\left\{\begin{array}{l}\cos(n-1)\alpha \\ +i\sin(n-1)\alpha\end{array}\right\}}{1 - 2c\cos\alpha + c^2}.$$

Hence, by equating real and imaginary parts, we have

$$C = \frac{1 - c\cos\alpha - c^n\cos n\alpha + c^{n+1}\cos(n-1)\alpha}{1 - 2c\cos\alpha + c^2},$$

and

$$S = \frac{c\sin\alpha - c^n\sin n\alpha + c^{n+1}\sin(n-1)\alpha}{1 - 2c\cos\alpha + c^2}.$$

The sum to infinity is obtained by omitting the terms containing c^n and c^{n+1}, which become indefinitely small when n is very great.

Hence

$$C_\infty = \frac{1 - c\cos\alpha}{1 - 2c\cos\alpha + c^2},$$

and

$$S_\infty = \frac{c\sin\alpha}{1 - 2c\cos\alpha + c^2}.$$

From the results for C and S it is now clear that the above series might have been summed, without the use of imaginary quantities, by multiplying both sides of (1) and (2) by the quantity $1 - 2c\cos\alpha + c^2$. The coefficients of c^3, c^2, ... c^{n-1} would then be found to vanish and the values of C and S be easily obtained.

■ 106. EXAMPLE: *Sum the series*

$$\frac{1}{2}\sin\alpha + \frac{1\cdot 3}{2\cdot 4}\sin 2\alpha + \frac{1\cdot 3\cdot 5}{2\cdot 4\cdot 6}\sin 3\alpha + ... \text{ad inf.}$$

Let

$$S = \frac{1}{2}\sin\alpha + \frac{1\cdot 3}{2\cdot 4}\sin 2\alpha + \frac{1\cdot 3\cdot 5}{2\cdot 4\cdot 6}\sin 3\alpha + ...,$$

and

$$C = 1 + \frac{1}{2}\cos\alpha + \frac{1\cdot 3}{2\cdot 4}\cos 2\alpha + \frac{1\cdot 3\cdot 5}{2\cdot 4\cdot 6}\cos 3\alpha + ...$$

Hence, multiplying the first by i and adding to the second, we have

$$C + Si = 1 + \frac{1}{2}e^{ai} + \frac{1\cdot 3}{2\cdot 4}e^{2ai} + \frac{1\cdot 3\cdot 5}{2\cdot 4\cdot 6}e^{3ai} + ...$$

$$= (1 - e^{ai})^{-1/2}, \text{ if } \alpha \neq 2n\pi,$$

by the Binomial Theorem. (Art. 26.)

$\therefore$

$$C + Si = \{1 - \cos\alpha - i\sin\alpha\}^{-1/2}$$

$$= \left\{2\sin\frac{\alpha}{2}\left(\sin\frac{\alpha}{2} - i\cos\frac{\alpha}{2}\right)\right\}^{-1/2}$$

$$= \left\{ 2\sin\frac{\alpha}{2} \right\}^{-1/2} \left\{ \cos\left(\frac{\alpha}{2} - \frac{\pi}{2}\right) + i\sin\left(\frac{\alpha}{2} - \frac{\pi}{2}\right) \right\}^{-1/2}$$

$$= \left\{ 2\sin\frac{\alpha}{2} \right\}^{-1/2} \left\{ \cos\left(\frac{\pi - \alpha}{4}\right) + i\sin\left(\frac{\pi - \alpha}{4}\right) \right\}.$$

Hence, by equating real and imaginary parts, we have

$$C = \left\{ 2\sin\frac{\alpha}{2} \right\}^{-1/2} \cos\frac{\pi - \alpha}{4},$$

and

$$S = \left\{ 2\sin\frac{\alpha}{2} \right\}^{-1/2} \sin\frac{\pi - \alpha}{4},$$

If $\qquad \alpha = 2n\pi$, clearly $S = 0$ and $C = \infty$.

EXAMPLES XVI

Sum the series

1. $\sin\alpha + \dfrac{1}{2}\sin 2\alpha + \dfrac{1}{2^2}\sin 3\alpha + \dots\dots$ ad inf.

2. $\cos\alpha \cdot \cos\alpha + \cos^2\alpha \cos 2\alpha + \cos^3\alpha \cos 3\alpha + \dots\dots$ ad inf.

3. $\sin\alpha \cdot \sin\alpha + \sin^2\alpha\sin 2\alpha + \sin^3\alpha\sin 3\alpha + \dots\dots$ ad inf., where $\alpha \ne \pm\dfrac{\pi}{2}$.

4. $\sin\alpha \cdot \cos\alpha + \sin^2\alpha \cdot \cos 2\alpha + \sin^2\alpha \cdot \cos 3\alpha + \dots\dots$ ad inf., where $\alpha \ne \pm\dfrac{\pi}{2}$.

5. $\sin\alpha + c\sin(\alpha + \beta) + c^2\sin(\alpha + 2\beta) + \dots\dots$ to n terms and ad inf.

6. $1 + c\cosh\alpha + c^2\cosh 2\alpha + \dots + c^{n-1}\cosh(n-1)\alpha.$

7. $c\sinh\alpha + c^2\sinh 2\alpha + \dots\dots + \dots\dots$ ad inf.

8. $1 - 2\cos\alpha + 3\cos 2\alpha - 4\cos 3\alpha + \dots\dots$ to n terms.

9. $3\sin\alpha + 5\sin 2\alpha + 7\sin 3\alpha + \dots\dots$ to n terms.

10. When $\alpha = \dfrac{\pi}{2}$, find what are the values of the series in Examples 3 and 4.

11. $\sin\alpha + n\sin(\alpha + \beta) + \dfrac{n(n-1)}{1\cdot 2}\sin(\alpha + 2\beta) + \dots\dots + $ to $(n + 1)$ terms, n being a positive integer.

12. $\sin\alpha + \dfrac{1}{2}\sin 3\alpha + \dfrac{1\cdot 3}{2\cdot 4}\sin 5\alpha + \dots\dots$ ad inf.

13. $\cos^n\alpha - n\cos^{n-1}\alpha\cos\alpha + \dfrac{n(n-1)}{1\cdot 2}\cos^{n-2}\alpha\cos 2\alpha + \dots + $ to $(n + 1)$ terms, n being a positive integer.

14. $n\sin\alpha + \dfrac{n(n+1)}{1\cdot 2}\sin 2\alpha + \dfrac{n(n+1)(n+2)}{1\cdot 2\cdot 3}\sin 3\alpha + \dots\dots$ ad inf.

15. $1 + \dfrac{1}{2}\cos 2\theta - \dfrac{1}{2 \cdot 4}\cos 4\theta + \dfrac{1 \cdot 3}{2 \cdot 4 \cdot 6}\cos 6\theta - \ldots\ldots$ ad inf.

16. $\sinh u + n \sinh 2u + \dfrac{n(n-1)}{1 \cdot 2}\sinh 3u + \ldots\ldots$ to $n+1$ terms, where n is a positive integer.

■ **107. EXAMPLE:** *Sum the series*

$$1 + \frac{c^2 \cos 2\theta}{\lfloor 2} + \frac{c^4 \cos 4\theta}{\lfloor 4} + \ldots\ldots \text{ad inf.}$$

Let $\qquad C \equiv 1 + \dfrac{c^2 \cos 2\theta}{\lfloor 2} + \dfrac{c^4 \cos 4\theta}{\lfloor 4} + \ldots\ldots \text{ad inf.}$ ... (1)

and $\qquad S \equiv \dfrac{c^2 \sin 2\theta}{\lfloor 2} + \dfrac{c^4 \sin 4\theta}{\lfloor 4} + \ldots\ldots \text{ad inf.}$ ... (2)

Hence

$$C + Si = 1 + \frac{c^2 e^{2\theta i}}{\lfloor 2} + \frac{c^4 e^{4\theta i}}{\lfloor 4} + \ldots. \text{ ad inf.}$$

$$= 1 + \frac{y^2}{\lfloor 2} + \frac{y^4}{\lfloor 4} + \frac{y^6}{\lfloor 6} + \ldots\ldots,$$

where $\qquad y = ce^{\theta i} = c(\cos\theta + i\sin\theta).$

$\therefore \qquad C + Si = \dfrac{e^y + e^{-y}}{2}$

$$= \frac{1}{2}e^{c\cos\theta + ie\sin\theta} + \frac{1}{2}e^{-c\cos\theta - ic\sin\theta} \qquad \ldots(3)$$

$$= \frac{1}{2}e^{-e\cos\theta}\left[\cos(c\sin\theta) + i\sin(c\sin\theta)\right]$$

$$+ \frac{1}{2}e^{-c\cos\theta}\left[\cos(c\sin\theta) - i\sin(c \text{ in }\theta)\right]. \qquad \text{(Art. 62)}$$

By equating real and imaginary parts we therefore have

$$C = \frac{1}{2}\cos(c\sin\theta)\left[e^{c\cos\theta} + e^{-c\cos\theta}\right]$$

$$= \cos(c\sin\theta)\cosh(c\cos\theta),$$

and $\qquad S = \dfrac{1}{2}\sin(c\sin\theta)\left[e^{c\cos\theta} - e^{-c\cos\theta}\right]$

$$= \sin(c\sin\theta)\sinh(c\cos\theta).$$

Aliter. From (3) we have

$$C + Si = \frac{1}{2}e^{(c\sin\theta - ic\cos\theta)i} + \frac{1}{2}e^{-(c\sin\theta - ic\cos\theta)i}$$

$$= \cos(c\sin\theta - ic\cos\theta) \qquad \text{(Art. 62)}$$

$$= [\cos (c \sin \theta) \cos (ic \cos \theta) + \sin (c \sin \theta) \sin (ic \cos \theta)]$$
$$= [\cos (c \sin \theta) \cosh (c \cos \theta) + i \sin (c \sin \theta) \sinh (c \cos \theta)] \qquad \text{(Art. 68)}$$

Hence C and S as before.

■ 108. EXAMPLE: *Sum the two series*

$$c \sin \alpha + \frac{c^2}{2} \sin 2\alpha + \frac{c^3}{3} \sin 3\alpha + \ldots\ldots \text{ ad inf.,}$$

and $\qquad c \cos \alpha + \dfrac{c^2}{2} \cos 2\alpha + \dfrac{c^3}{3} \cos 3\alpha + \ldots\ldots \text{ ad inf.,}$

where c is numerically not greater than unity.

Let S and C stand for these two series; then, as before, we have

$$C + Si = c(\cos\alpha + i\sin\alpha) + \frac{c^2}{2}(\cos 2\alpha + i\sin 2\alpha) + \ldots\ldots$$

$$= ce^{\alpha i} + \frac{c^2}{2}e^{2\alpha i} + \frac{c^3}{3}e^{3\alpha i} + \ldots\ldots \qquad \ldots(1)$$

$$= - \log [1 - ce^{\alpha i}] \text{ (by Art. 90)} \qquad \ldots(2)$$

$$= - \log [1 - c \cos \alpha - ic \sin \alpha] \qquad \text{(Art. 62)}$$

Let $1 - c \cos \alpha = r \cos \theta$, and $-c \sin \alpha = r \sin \theta$, so that

$$r = +\sqrt{1 - 2c\cos\alpha + c^2}, \quad \cos\theta = \frac{1 - c\cos\alpha}{r},$$

and

$$\sin\theta = -\frac{c\sin\alpha}{r}, \quad \textit{i.e.,} \quad \theta = \tan^{-1}\frac{-c\sin\alpha}{1 - c\cos\alpha},$$

with the convention of Art. 20.

$$\therefore \qquad C + Si = -\log\left[\sqrt{1 - 2c\cos\alpha + c^2}\ (\cos\theta + i\sin\theta)\right]$$

$$= -\log\left[\sqrt{1 - 2c\cos\alpha + c^2} \cdot e^{\theta i}\right]$$

$$= -\log\sqrt{1 - 2c\cos\alpha + c^2} - \theta i.$$

$$\therefore \qquad C = -\log\sqrt{(1 - 2c\cos\alpha + c^2)} = -\frac{1}{2}\log(1 - 2c\cos\alpha + c^2) \qquad \ldots(3)$$

and $\qquad S = -\theta = -\tan^{-1}\left(\dfrac{-c\sin\alpha}{1 - c\cos\alpha}\right) \qquad \ldots(4)$

Exceptional cases. When $c = 1$, the quantity (2)

$$= - \log [1 - \cos \alpha - i \sin \alpha] = - \log [1 + \cos (\alpha - \pi) + i \sin (\alpha - \pi)].$$

This, by Art. 90, is always equal to the series (1) except when $\alpha - \pi$ is equal to $(2n + 1)\pi$, *i.e.*, except when α is a multiple of 2π.

In this case $S = 0$,

and $\qquad C = +1 + \dfrac{1}{2} + \dfrac{1}{3} + \dfrac{1}{4} + \ldots\ldots,$

which is known to be a divergent series.

When $c = 1$, the quantity (2)

$$= -\log[1 + \cos\alpha + i\sin\alpha].$$

This by Art. 90 is always equal to the series (1) except when $\alpha = (2n + 1)\pi$. In this case $S = 0$, and

$$C = 1 + \frac{1}{2} + \frac{1}{3} + \frac{1}{4} + \ldots$$

The results (3) and (4) give then the sum of the two series except when (1) $c = 1$ and $\alpha = 2n\pi$, (2) $c = -1$ and $\alpha = (2n + 1)\pi$, and (3) when $c > 1$.

In examples depending on the logarithm series it will be often found that for some particular values of the angle there is no sum.

Particular case. Let $c = \cos\alpha$, where α lies between 0 and $\dfrac{\pi}{2}$, so that

$$S = \cos\alpha \cdot \sin\alpha + \frac{1}{2}\cos^2\alpha\sin 2\alpha + \frac{1}{3}\cos^3\alpha\sin 3\alpha + \ldots$$

In this case

$$S = -\tan^{-1}\left(\frac{-\sin\alpha\cos\alpha}{\sin^2\alpha}\right), \text{ by (4)},$$

$$= -\tan^{-1}(-\cot\alpha)$$

$$= -\left(\alpha - \frac{\pi}{2}\right),$$

remembering the convention mentioned above,

$$= \frac{\pi}{2} - \alpha.$$

EXAMPLES XVII

Sum the series

1. $\sin\alpha + c\sin(\alpha + \beta) + \dfrac{c^2}{\lfloor 2}\sin(\alpha + 2\beta) + \ldots$ ad inf.

2. $\cos\alpha + c\cos(\alpha + \beta) + \dfrac{c^2}{\lfloor 2}\cos(\alpha + 2\beta) + \ldots$ ad inf.

3. $1 - \cos\alpha\cos\beta + \dfrac{\cos^2\alpha}{\lfloor 2}\cos 2\beta - \dfrac{\cos^3\alpha}{\lfloor 3}\cos 3\beta + \ldots$ ad inf.

4. $\sin\alpha - \dfrac{\sin(\alpha + 2\beta)}{\lfloor 2} + \dfrac{\sin(\alpha + 4\beta)}{\lfloor 4} - \ldots$ ad inf.

5. $\cos\alpha - \dfrac{\cos(\alpha + 2\beta)}{\lfloor 3} + \dfrac{\cos(\alpha + 4\beta)}{\lfloor 5} - \ldots$ ad inf.

6. $1+\cosh\alpha+\dfrac{\cosh 2\alpha}{\underline{|2}}+\dfrac{\cosh 3\alpha}{\underline{|3}}+\ldots$ ad inf.

7. $\sinh\alpha+\dfrac{\sinh 2\alpha}{\underline{|2}}+\dfrac{\sinh 3\alpha}{\underline{|3}}+\ldots$ ad inf.

8. $1+e^{\cos\alpha}\cos(\sin\alpha)+\dfrac{1}{\underline{|2}}e^{2\cos\alpha}\cdot\cos(2\sin\alpha)+\ldots$ ad inf.

9. $1+e^{\sin\alpha}\cos(\cos\alpha)+\dfrac{e^{2\sin\alpha}}{\underline{|2}}\cos(2\cos\alpha)+\dfrac{e^{3\sin\alpha}}{\underline{|3}}\cos(3\cos\alpha)+\ldots$ ad inf.

10. $\dfrac{5\cos\theta}{\underline{|1}}+\dfrac{7\cos 3\theta}{\underline{|3}}+\dfrac{9\cos 5\theta}{\underline{|5}}+\ldots$ ad inf.

[In the following examples c may be assumed to be positive and not greater than unity; when c equals unity there will be, as in Art. 108, exceptional cases for some values of the angle α.]

11. $c\sin\alpha-\dfrac{c^2}{2}\sin 2\alpha+\dfrac{c^3}{3}\sin 3\alpha-\ldots$ ad inf.

12. $c\sin\alpha+\dfrac{1}{3}c^3\sin 3\alpha+\dfrac{1}{5}c^5\sin 5\alpha+\ldots$ ad inf.

13. $c\cos\alpha+\dfrac{1}{3}c^2\cos 3\alpha+\dfrac{1}{5}c^5\cos 5\alpha+\ldots$ ad inf.

14. $c\cos\alpha-\dfrac{1}{3}c^2\cos 3\alpha+\dfrac{1}{5}c^5\cos 5\alpha-\ldots$ ad inf.

15. $c\sin\alpha-\dfrac{1}{3}c^3\sin 3\alpha+\dfrac{1}{5}c^5\sin 5\alpha-\ldots$ ad inf.

16. $\cos\alpha-\dfrac{1}{3}\cos 3\alpha+\dfrac{1}{5}\cos 5\alpha-\ldots$ ad inf.

17. $c\cos\alpha-\dfrac{1}{3}c^3\cos(\alpha+2\beta)+\dfrac{1}{5}c^5\cos(\alpha+4\beta)-\ldots$ ad inf.

18. $\sin\alpha\sin\beta+\dfrac{1}{2}\sin 2\alpha\sin 2\beta+\dfrac{1}{3}\sin 3\alpha\sin 3\beta+\ldots$ ad inf.

19. $c\sin^2\alpha-\dfrac{1}{2}c^2\sin^2 2\alpha+\dfrac{1}{3}c^3\sin^2 3\alpha-\ldots$ ad inf.

20. $\sinh\alpha-\dfrac{1}{\pi}\sinh 2\alpha+\dfrac{1}{\pi}\sinh 3\alpha-\ldots$ ad inf.

21. $e^{\alpha}\cos\beta-\dfrac{1}{3}e^{3\alpha}\cos 3\beta+\dfrac{1}{5}e^{2\alpha}\cos 5\beta-\ldots$ ad inf.

22. $\cos\dfrac{\pi}{3}+\dfrac{1}{3}\cos\dfrac{2\pi}{3}+\dfrac{1}{5}\cos\dfrac{3\pi}{3}+\dfrac{1}{7}\cos\dfrac{4\pi}{3}+\ldots$ ad inf.

23. If $\theta-\alpha=\tan^2\dfrac{\omega}{2}\sin 2\theta-\dfrac{1}{2}\tan^4\dfrac{\omega}{2}\sin 4\theta+\dfrac{1}{3}\tan^6\dfrac{\omega}{2}\sin 6\theta-\ldots$ ad inf.

prove that $\tan\alpha=\tan\theta\cdot\cos\omega$

24. If θ and ϕ be positive acute angles prove that the sum of the series.

$$\sin\theta\cos\phi + \frac{1}{3}\sin 3\theta\cos 3\phi + \frac{1}{5}\sin 5\theta\cos 5\phi + \dots \text{ ad inf.}$$

is $\dfrac{\pi}{4}$ or 0, according as $\theta >$ or $< \phi$.

Prove that

25. $\tanh x + \dfrac{1}{3}\tanh^3 x + \dfrac{1}{5}\tanh^5 x + \dots$

$$= \tan x - \frac{1}{3}\tan^3 x + \frac{1}{5}\tan^5 x - \dots, \text{ where } x \text{ lies between } -\frac{\pi}{4} \text{ and } +\frac{\pi}{4}.$$

26. $2\sin^2\theta + \dfrac{1}{2}\cdot 4\sin^4\theta + \dfrac{1}{3}\cdot 8\sin^6\theta + \dots$

$$= 2\left(\tan^2\theta + \frac{1}{3}\tan^6\theta + \frac{1}{5}\tan^{10}\theta + \dots\right), \text{ where } \theta \text{ lies between } -\frac{\pi}{4} \text{ and } +\frac{\pi}{4}.$$

27. $\sin\theta + \dfrac{1}{3}\sin^3\theta + \dfrac{1}{5}\sin^5\theta + \dots$

$$= 2\left(\sin\theta - \frac{1}{3}\sin 3\theta + \frac{1}{5}\sin 5\theta - \dots\right), \text{ where } \theta \neq (2n+1)\frac{\pi}{2}.$$

■ **109.** We subjoin some examples of series which come under neither of the foregoing heads nor under that of Chapter XIX, Part I. In general they are to be summed by the artifice of splitting each term into the difference of two terms. Considerable ingenuity is often required. When the answer is known the method of summation can usually be easily seen; for the answer when n is put equal to unity gives the form in which the first term of the series has to be put.

■ **EXAMPLE 1:** *Sum to n terms the series*

$$\sin^3\frac{\theta}{3} + 3\sin^3\frac{\theta}{3^2} + 3^2\sin^3\frac{\theta}{3^3} + \dots$$

Since always $\sin 3\phi = 3\sin\phi - 4\sin^3\theta$, we have

$$\sin^2\frac{\theta}{3} = \frac{1}{4}\left(3\sin\frac{\theta}{3} - \sin\theta\right),$$

$$3\cdot\sin^3\frac{\theta}{3^2} = \frac{1}{4}\cdot 3\cdot\left[3\sin\frac{\theta}{3^2} - \sin\frac{\theta}{3}\right] = \frac{1}{4}\left[3^2\sin\frac{\theta}{3^2} - 3\sin\frac{\theta}{3}\right],$$

$$3^2\sin^2\frac{\theta}{3^2} = \frac{1}{4}\left[3^3\sin\frac{\theta}{3^2} - 3^2\sin\frac{\theta}{3^2}\right],$$

. . . .

$$3^{n-1}\sin^2\frac{\theta}{3^n} = \frac{1}{4}\left[3^n\sin\frac{\theta}{3^n} - 3^{n-1}\sin\frac{\theta}{3^{n-1}}\right].$$

Hence, by addition, the required sum

$$= \frac{1}{4}\left[3^n \sin\frac{\theta}{3^n} - \sin\theta \right].$$

Also the sum to infinity

$$= \frac{1}{4}[\theta - \sin\theta].$$ (Art. 228, Part I.)

▌ **EXAMPLE 2:** *Sum the series*

$$\tan\alpha + 2\tan 2\alpha + 2^2 \tan 2^2\alpha + \ldots + 2^{n-1}\tan 2^{n-1}\alpha.$$

We have easily

$$\tan\alpha = \cot\alpha - 2\cot 2\alpha,$$

$$\tan 2\alpha = \cot 2\alpha - 2\cot 2^2\alpha,$$

$$\tan 2^2\alpha = \cot 2^2\alpha - 2\cot 2^2\alpha,$$

and $$\tan 2^{n-1}\alpha = \cot 2^{n-1}\alpha - 2\cot 2^n\alpha.$$

By multiplying these rows in succession by 1, 2, 2^2, ... 2^{n-1} we have $\tan\alpha + 2\tan 2\alpha + 2^2\tan 2^2\alpha + \ldots + 2^{n-1}\tan 2^{n-1}\alpha = \cot\alpha - 2^n\cot 2^n\alpha,$

the other terms all disappearing.

The required sum therefore $= \cot\alpha - 2^n\cot 2^n\alpha.$

▌ **EXAMPLE 3:** *Sum the series*

$$\tan\alpha\tan(\alpha + \beta) + \tan(\alpha + \beta)\tan(\alpha + 2\beta) + \tan(\alpha + 2\beta)\tan(\alpha + 3\beta) + \ldots\ldots$$
                                                    to *n* terms.

Let $u_r \equiv$ the *r*th term, *i.e.*,

$$\tan\{\alpha + (r - 1)\beta\}\tan\{\alpha + r\beta\},$$

∴ $$(u_r + 1)\tan\beta$$

$$= \left[1 + \tan\{\alpha + (r-1)\beta\}\tan(\alpha + r\beta)\right] \times \tan\left[\alpha + r\beta - (\overline{\alpha + r - 1}\beta)\right]$$

$$= \tan\{\alpha + r\beta\} - \tan\{\alpha + \overline{r - 1}\beta\}.$$ [Art. 98, Part I.]

Hence giving *r* in succession the values 1, 2, ... *n*, we have

$$(1 + u_1)\tan\beta = \tan(\alpha + \beta) - \tan\alpha,$$

$$(1 + u_2)\tan\beta = \tan(\alpha + 2\beta) - \tan(\alpha + \beta),$$

. . . .

$$(1 + u_n)\tan\beta = \tan(\alpha + n\beta) - \tan\{\alpha + (n - 1)\beta\}.$$

Hence, by addition,

$$(n + S_n)\tan\beta = \tan(\alpha + n\beta) - \tan\alpha,$$

so that $$S_n = \frac{\tan(\alpha + n\beta) - \tan\alpha - n\tan\beta}{\tan\beta}.$$

EXAMPLES XVIII

Sum the series

1. $\operatorname{cosec} \theta + \operatorname{cosec} 2\theta + \operatorname{cosec} 4\theta + \dots$ to n terms.

2. $\operatorname{cosec} \theta \operatorname{cosec} 2\theta + \operatorname{cosec} 2\theta \operatorname{cosec} 3\theta + \operatorname{cosec} 3\theta \operatorname{cosec} 4\theta + \dots$ to n terms.

3. $\sec \theta \sec 2\theta + \sec 2\theta \sec 3\theta + \sec 3\theta \sec 4\theta + \dots$ to n terms.

4. $\sec \theta \sec (\theta + \phi) + \sec (\theta + \phi)\sec (\theta + 2\phi) + \sec (\theta + 2\phi)\sec (\theta + 3\phi) + \dots$ to n terms.

5. $\dfrac{1}{\cos \alpha + \cos 3\alpha} + \dfrac{1}{\cos \alpha + \cos 5\alpha} + \dfrac{1}{\cos \alpha + \cos 7\alpha} + \dots$ to n terms.

6. $\tan \theta + \dfrac{1}{2}\tan\dfrac{\theta}{2} + \dfrac{1}{2^2}\tan\dfrac{\theta}{2^2} + \dfrac{1}{2^3}\tan\dfrac{\theta}{2^3} + \dots$ ad inf.

7. $\tanh \theta + \dfrac{1}{2}\tanh\dfrac{\theta}{2} + \dfrac{1}{2^2}\tanh\dfrac{\theta}{2^2} + \dfrac{1}{2^3}\tanh\dfrac{\theta}{2^3} + \dots$ to n terms.

8. $\tan \theta \sec 2\theta + \tan 2\theta \sec 4\theta + \tan 4\theta \sec 8\theta + \dots$ to n terms.

9. $\tan\dfrac{\theta}{2}\sec\theta + \tan\dfrac{\theta}{2^2}\sec\dfrac{\theta}{2} + \tan\dfrac{\theta}{2^3}\sec\dfrac{\theta}{2^2} + \dots$ to n terms and to infinity.

10. $\dfrac{1}{2\cos\theta} + \dfrac{1}{2^2\cos\theta\cos 2\theta} + \dfrac{1}{2^3\cos\theta\cos 2\theta\cos 2^2\theta} + \dots$ to n terms.

11. $\sin 2\theta\cos^2\theta - \dfrac{1}{2}\sin 4\theta\cos^2 2\theta + \dfrac{1}{4}\sin 8\theta\cos^2 4\theta - \dots$ to n terms.

12. $\sin 2\theta\sin^2\theta + \dfrac{1}{2}\sin 4\theta\cos^2 2\theta + \dfrac{1}{4}\sin 8\theta\sin^2 4\theta + \dots$ to n terms.

13. $\dfrac{\sin\theta}{\cos\theta + \cos 2\theta} + \dfrac{\sin 2\theta}{\cos\theta + \cos 4\theta} + \dfrac{\sin 3\theta}{\cos\theta + \cos 6\theta} + \dots$ to n terms.

14. $\tan^2\alpha\tan 2\alpha + \dfrac{1}{2}\tan^2 2\alpha\tan 4\alpha + \dfrac{1}{2^2}\tan^2 4\alpha\tan 8\alpha + \dots$ to n terms.

15. $\cos^3\theta - \dfrac{1}{3}\cos^3 3\theta + \dfrac{1}{3^2}\cos^3 3^2\theta - \dfrac{1}{3^3}\cos^3 3^3\theta + \dots$ to n terms.

16. $\sin^3\dfrac{\theta}{3} + 3\sin^3\dfrac{\theta}{3^2} + 3^2\sin^3\dfrac{\theta}{3^3} + \dots$ to n terms.

17. $\dfrac{1}{\cot\theta - 3\tan\theta} + \dfrac{3}{\cot 3\theta - 3\tan 3\theta} + \dfrac{3^2}{\cot 3^2\theta - 3\tan 3^2\theta} + \dots$ to n terms.

18. $\dfrac{\cos\theta - \cos 3\theta}{\sin 3\theta} + 3\dfrac{\cos 3\theta - \cos 3^2\theta}{\sin 3^2\theta} + 3^2\dfrac{\cos 3^2\theta - \cos 3^3\theta}{\sin 3^2\theta} + \dots$ to n terms.

19. $\tan^{-1}\dfrac{4}{1+3.4} + \tan^{-1}\dfrac{6}{1+8.9} + \tan^{-1}\dfrac{8}{1+15.16} + \dots$ to n terms.

20. $\tan^{-1}\dfrac{1}{3} + \tan^{-1}\dfrac{1}{7} + \tan^{-1}\dfrac{1}{13} + \tan^{-1}\dfrac{1}{21} + \dots\dots$ to n terms.

21. $\tan^{-1}\dfrac{1}{3} + \tan^{-1}\dfrac{2}{9} + \dots\dots + \tan^{-1}\dfrac{2^{n-1}}{1+2^{2n-1}} + \dots\dots$ ad inf.

22. $\sin^{-1}\dfrac{1}{\sqrt{2}} + \sin^{-1}\dfrac{\sqrt{2}-1}{\sqrt{6}} + \sin^{-1}\dfrac{\sqrt{3}-\sqrt{2}}{\sqrt{12}} + \dots\dots + \sin^{-1}\dfrac{\sqrt{n}-\sqrt{n-1}}{\sqrt{n(n+1)}} + \dots\dots$ ad inf.

Expansions

▪ **110.** In some branches of higher Mathematics it is desirable to be able to expand certain quantities in a series of ascending powers.

As an example we will expand

$$\log(1 - 2\alpha\cos\theta + a^2)$$

in ascending powers a.

Since $\qquad 2\cos\theta = e^{\theta i} + e^{-\theta i},$

we have

$\log(1 - 2a\cos\theta + \alpha^2) = \log[1 - a(e^{\theta i} + e^{-\theta i}) + a^2]$

$\qquad = \log[(1 - ae^{\theta i})(1 - ae^{-\theta i})]$

$\qquad = \log[(1 - ae^{\theta i}) + \log(1 - ae^{-\theta i})]$

$= -ae^{\theta i} - \dfrac{1}{2}a^2 e^{2\theta i} - \dfrac{1}{3}a^3 e^{3\theta i} - \dfrac{1}{4}a^4 e^{4\theta i} - ae^{-\theta i} - \dfrac{1}{2}a^2 e^{-2\theta i} - \dfrac{1}{3}a^3 e^{-3\theta i} - \dots\dots$

$= -a[e^{\theta i} + e^{-\theta i}] - \dfrac{1}{2}a^2[e^{2\theta i} + e^{-2\theta i}] - \dfrac{1}{3}a^3[e^{3\theta i} + e^{-3\theta i}] - \dots\dots$

$= -a \cdot 2\cos\theta - \dfrac{1}{2}a^2 \cdot 2\cos 2\theta - \dfrac{1}{3}a^3 \cdot 2\cos 3\theta \dots\dots$

$= -2\left[a\cos\theta + \dfrac{1}{2}a^2\cos 2\theta + \dfrac{1}{3}a^3\cos 3\theta + \dots\dots\right].$

The expansion of $\log(1 - ae^{\theta i})$ is legitimate, by Art. 90, if the modulus of $-ae^{\theta i}$ be less than unity.

Now $\qquad -ae^{\theta i} = a\{\cos(\pi + \theta) + i\sin(\pi + \theta)\},$

so that its modulus is equal to a. Hence the above expansion is legitimate provided that a is less than unity.

The expansion is also legitimate if a be equal to unity, provided that θ do not equal an even multiple of π.

It is also legitimate if a be equal to -1 and θ do not equal an odd multiple of π.

▎**111. EXAMPLE:** *Expand*

$$\frac{1-a^2}{1-2a\cos\theta + a^2}$$

in a series of ascending powers of a.

We have

$$\frac{1-a^2}{1-2a\cos\theta+a^2} = -1 + \frac{2-2a\cos\theta}{1-2a\cos\theta+a^2}$$

$$= -1 + \frac{2-a(e^{\theta i}+e^{-\theta i})}{1-a(e^{\theta i}+e^{-\theta i})+a^2}$$

$$= -1 + \frac{2-a(e^{\theta i}+e^{-\theta i})}{(1-ae^{\theta i})(1-ae^{-\theta i})}$$

$$= -1 + \frac{1}{1-ae^{\theta i}} + \frac{1}{1-ae^{-\theta i}}$$

$$= -1 + (1-ae^{\theta i})^{-1} + (1-ae^{-\theta i})^{-1}$$

$$= -1 + 1 + ae^{\theta i} + a^2 e^{2\theta i} + a^3 e^{3\theta i} + \ldots\ldots$$

$$\qquad + 1 + ae^{-\theta i} + a^2 e^{-2\theta i} + a^3 e^{-3\theta i} + \ldots\ldots$$

$$= 1 + a(e^{\theta i}+e^{-\theta i}) + a^2(e^{2\theta i}+e^{-2\theta i}) + \ldots\ldots$$

$$= 1 + 2a\cos\theta + 2a^2\cos 2\theta + 2a^3\cos 3\theta + \ldots \text{ ad inf.}$$

The expansions of $(1-ae^{\theta i})^{-1}$ and $(1-ae^{-\theta i})^{-1}$ by the Binomial Theorem are legitimate if the modulus of $ae^{\theta i}$ be less than unity, *i.e.*, if a be numerically < 1, but not otherwise. (Art. 26).

The above series is the one assumed in Art. 49.

Similarly we can deduce the series of Art. 48. For we have

$$\frac{2a\sin\theta}{1-2a\cos\theta+a^2} = \frac{1}{i}\frac{a(e^{\theta i}-e^{-\theta i})}{1-a(e^{\theta i}+e^{-\theta i})+a^2}$$

$$= \frac{1}{i}\frac{ae^{\theta i}-ae^{-\theta i}}{(1-ae^{\theta i})(1-ae^{-\theta i})} = \frac{1}{i}\left[\frac{1}{1-ae^{\theta i}} - \frac{1}{1-ae^{-\theta i}}\right]$$

$$= \frac{1}{i}\left\{(1+ae^{\theta i}+a^2 e^{2\theta i}+\ldots) - (1+ae^{-\theta i}+a^2 e^{-2\theta i}+\ldots)\right\}$$

$$= 2a\sin\theta + 2a^2\sin 2\theta + 2a^3\sin 3\theta + \ldots \text{ ad inf.}$$

As before this expansion is legitimate only if $a < 1$.

■ **112. EXAMPLE:** *If* $\sin x = n\sin(\alpha+x)$, *expand* x *in a series of ascending powers of* n, *where* n *is less than unity.*

Since

$$\sin x = n(\sin(\alpha+x) = n(\sin\alpha\cos x + \cos\alpha\sin x),$$

$$\therefore \qquad \tan x = \frac{n\sin\alpha}{1-n\cos\alpha},$$

$$\therefore \qquad \frac{e^{xi}-e^{-xi}}{e^{xi}+e^{-xi}} = \frac{ni\sin\alpha}{1-n\cos\alpha},$$

$$\therefore \qquad \frac{e^{xi}}{e^{-xi}} = \frac{1-n\cos\alpha+ni\sin\alpha}{1-n\cos\alpha-ni\sin\alpha} = \frac{1-ne^{-\alpha i}}{1-ne^{\alpha i}},$$

$\therefore \qquad 2xi = \log(1 - ne^{-ai}) - \log(1 - ne^{ai})$

$$= -ne^{-ai} - \frac{1}{2}n^2 e^{-2ai} - \frac{1}{3}n^3 e^{-3ai} - \ldots$$

$$+ne^{ai} + \frac{1}{2}n^2 e^{2ai} + \frac{1}{3}n^3 e^{3ai} + \ldots$$

$$= n(e^{ai} - e^{-ai}) + \frac{1}{2}n^2 (e^{2ai} - e^{-2ai})$$

$$+ \frac{1}{3}n^3 (e^{3ai} - e^{-3ai}) \ldots \text{ ad inf.}$$

$$= n \cdot 2i\sin\alpha + \frac{1}{2}n^2 \cdot 2i\sin 2\alpha + \frac{1}{3}n^3 \cdot 2i\sin 3\alpha + \ldots$$

$\therefore \qquad x = n\sin\alpha + \frac{1}{2}n^2 \sin 2\alpha + \frac{1}{3}n^3 \sin 3\alpha + \ldots \qquad \ldots(1)$

In this equation we have assumed x to lie between $-\frac{\pi}{2}$ and $+\frac{\pi}{2}$; if it do not, then, instead of $2xi$, we should read $2k\pi i + 2xi$; the left hand of equation (1) would then be $x + k\pi$, and we must choose k so that $x + k\pi$ shall lie between $-\frac{\pi}{2}$ and $+\frac{\pi}{2}$.

As before the expansions are legitimate if n be < unity.

▮ **113. EXAMPLE:** *Expand $e^{ax} \cos bx$ in a series of ascending powers of x.*
We have

$$e^{ax} \cos bx = e^{ax} \cdot \frac{e^{bxi} + e^{-bxi}}{2}$$

$$= \frac{1}{2}e^{(a+bi)x} + \frac{1}{2}e^{(a-bi)x}$$

$$= \frac{1}{2}\left[1 + (a+bi)x + \frac{(a+bi)^2 x^2}{\lfloor 2} + \frac{(a+bi)^3 x^3}{\lfloor 3} + \ldots\right]$$

$$+ \frac{1}{2}\left[1 + (a-bi)x + \frac{(a-bi)^2 x^2}{\lfloor 2} + \ldots\right]$$

The coefficient of x^n

$$= \frac{(a+bi)^n + (a-bi)^n}{2\lfloor n}$$

If $a + bi = r \cos(a + i \sin \alpha)$, so that

$$r = +\sqrt{a^2 + b^2} \text{ and } \tan\alpha = \frac{b}{a},$$

with the convention of Art. 20, then the coefficient of x^n

$$= \frac{\{r(\cos\alpha + i\sin\alpha)\}^n + \{r(\cos\alpha - i\sin\alpha)\}^n}{2\lfloor n}$$

$$= r^n \frac{\cos n\alpha}{\lfloor n},$$

by De Moivre's Theorem.

Hence we have

$$e^{\alpha x}\cos bx = 1 + r\cos\alpha \cdot x + \frac{r^2\cos 2\alpha}{\lfloor 2}x^2 + \frac{r^3\cos 3\alpha}{\lfloor 3}x^3 + ...$$

where

$$r = +\sqrt{a^2 + b^2} \text{ and } \tan\alpha = \frac{b}{a}.$$

This expansion is legitimate for all values of a, b and x (Art. 57.)

EXAMPLES XIX

Expand in an infinite series.

1. $\dfrac{1 + a\cos\theta}{1 + 2a\cos\theta + a^2}$

2. $\dfrac{\cos\theta - a\cos(\theta - \phi)}{1 - 2a\cos\phi + a^2}$

3. $\dfrac{\sin\theta - a\sin(\theta - \phi)}{1 - 2a\cos\phi + a^2}$

4. $e^{a\cos\phi}\cos(\theta + a\sin\phi)$

5. $e^{e\theta}\sin b\theta$.

Prove that

6. $\log\dfrac{a^2}{a^2\cos^2\theta + b^2\sin^2\theta} = 4\left(c\sin^2\theta - \dfrac{1}{2}c^2\sin^2 2\theta + \dfrac{1}{3}c^3\sin^2 3\theta - ...\right),$

where $c = \dfrac{a-b}{a+b}$.

7. $\tan^{-1}\dfrac{a\sin\theta}{1 - a\cos\theta} = a\sin\theta + \dfrac{1}{2}a^2\sin 2\theta + \dfrac{1}{3}a^2\sin 3\theta + ...$ ad inf.

8. $\dfrac{1}{2}\tan^{-1}(\sin\alpha\tan 2\beta) = \sin\alpha\tan\beta + \dfrac{1}{3}\sin 3\alpha\tan^2\beta + \dfrac{1}{5}\sin 5\alpha\tan^5\beta + ...$ ad inf.

9. If $\sin\theta = x\cos(\theta + \alpha)$, expand θ in a series of ascending powers of x.

10. Expand y in terms of $\cos\alpha$, where

$$2\tan y = \sin x\cosec\frac{x+\alpha}{2}\cosec\frac{x-\alpha}{2}.$$

11. If $\tan x = n\tan y$, and $m = \dfrac{1-n}{1+n}$, prove that

$$x + r\pi = y - m\sin 2y + \frac{m^2}{2}\sin 4y - \frac{m^3}{3}\sin 6y + \ldots \text{ ad inf.},$$

where r is to be so chosen that $x + r\pi - y$ lies between $-\frac{\pi}{2}$ and $+\frac{\pi}{2}$.

12. What does the series of the preceding question become when (1) $n = \cos a$, and (2)
$n = \frac{1}{\cos 2a}$?

13. Expand $\log \cos\left(\frac{\pi}{4} + \theta\right)$ in a series of sines and cosines of ascending multiples of θ.

14. Expand $\log \tan\left(\frac{\pi}{4} + \frac{\theta}{2}\right)$ in a series of sines of ascending multiples of θ.

15. Prove that
$$(1 + e^{\theta i}\tan\alpha)(1 + e^{-\theta i}\tan\alpha)(1 + e^{\theta i}\cot\alpha)(1 + e^{-\theta i}\cot\alpha) = 4\,(\sec\beta + \cos\theta)^2,$$

where $\beta = \frac{\pi}{2} - 2\alpha$.

Hence expand $\log(1 + \cos\beta\cos\theta)$ in a series of cosines of multiples of θ.

16. Prove that
$$\frac{2a\cos\theta}{1 - 2a\sin\theta + a^2} = 2a\cos\theta + 2a^2\sin 2\theta - 2a^3\cos 3\theta - 2a^4\sin 4\theta + \ldots \text{ ad inf.}$$

17. Prove that
$$\log\cos\theta = -\log 2 + \cos 2\theta - \frac{1}{2}\cos 4\theta + \frac{1}{3}\cos 6\theta - \ldots \text{ ad inf.,}$$

if θ be an angle whose cosine is positive.

18. In any triangle where $a > b$, prove that
$$\log c = \log a - \frac{b}{a}\cos C - \frac{1}{2}\frac{b^2}{a^2}\cos 2C - \frac{1}{3}\frac{b^3}{a^3}\cos 3C - \ldots \text{ ad inf.}$$

$$\left[\text{We have } c^2 = a^2 + b^2 - 2ab\cos C = a^2\left(1 - \frac{b}{a}e^{iC}\right)\left(1 - \frac{b}{a}e^{-iC}\right).\right]$$

19. Prove that the coefficient of x^n is the expansion of
$$e^{ax}\sin bx + e^{bx}\sin ax$$
in powers of x is
$$\frac{2(a^2 + b^2)^{n/2}}{\underline{|n}}\sin\frac{n\pi}{4}\cos\frac{n}{2}\left[\frac{\pi}{2} - 2\tan^{-1}\frac{b}{a}\right].$$

20. Prove that the coefficient of c^n in the expansion of
$$\log(a^3 + b^3 + c^3 - 3abc)$$
is $\frac{1}{n}\left[\frac{(-1)^{n-1}}{(a+b)^n} - \frac{2\cos n\theta}{(a^2 + b^2 - ab)^{n/2}}\right],$

where $\tan\theta = \frac{a-b}{a+b}\sqrt{3}.$

Resolution into Factors
Infinite Products for Sin θ and Cos θ

■ **114.** We know from Algebra that, if P be any expression containing x and if the value $x = \alpha$ would make P vanish, then $x - \alpha$ is a factor of P.

Hence to find the factors of any expression P we first solve the equation $P = O$. Also if P be of the nth degree we know that there are only n solutions of the equation $P = 0$. If the roots thus found are α, β, γ, ... κ, we know that $x - \alpha$, $x - \beta$, ... $x - \kappa$ are factors of the expression P and that there are no other factors which contain x.

We shall apply this method in the following articles.

■ **115.** *To resolve into factors the expression*

$$x^{2n} - 2x^n \cos n\theta + 1.$$

We have first to solve the equation

$$x^{2n} - 2x^n \cos n\theta + 1 = 0,$$

i.e., $$x^{2n} - 2x^n \cos n\theta + \cos^2 n\theta = -\sin^2 n\theta,$$

so that $$x^n - \cos n\theta = \pm\sqrt{-1}\,\sin n\theta,$$

and therefore

$$x = [\cos n\theta \pm \sqrt{-1}\,\sin n\theta]^{1/n}.$$

As in Art. 24 the values of this expression are the $2n$ quantities

$$\cos\theta \pm i\sin\theta,\; \cos\left(\theta + \frac{2\pi}{n}\right) \pm i\sin\left(\theta + \frac{2\pi}{n}\right),$$

$$\cos\left(\theta + \frac{4\pi}{n}\right) \pm i\sin\left(\theta + \frac{4\pi}{n}\right), \ldots\ldots$$

$$\cos\left\{\theta + \frac{2(n-1)\pi}{n}\right\} \pm i\sin\left\{\theta + \frac{2(n-1)\pi}{n}\right\}.$$

Taking the first pair of these quantities we have the corresponding factors

$$x - \cos\theta - i\sin\theta \text{ and } x - \cos\theta + i\sin\theta,$$

or, in one factor,

$$(x - \cos\theta)^2 + \sin^2\theta,$$

i.e., the quadratic factor

$$x^2 - 2x\cos\theta + 1.$$

104

Similarly the second, third,... pairs of the above quantities give as factors respectively

$$x^2 - 2x\cos\left(\theta + \frac{2\pi}{n}\right) + 1,$$

$$x^2 - 2x\cos\left(\theta + \frac{4\pi}{n}\right) + 1,$$

. . . . .

and
$$x^2 - 2x\cos\left\{\theta + \frac{2n-2}{n}\pi\right\} + 1.$$

Also on multiplying together these n factors we see that the coefficient of x^{2n} in their product is unity, which is also the coefficient of x^{2n} in the original expression. No other numerical factor is therefore required.

Hence,

$$x^{2n} - 2x^n \cos n\theta + 1$$

$$= \{x^2 - 2x\cos\theta + 1\}\left\{x^2 - 2x\cos\left(\theta + \frac{2\pi}{n}\right) + 1\right\}$$

$$\left\{x^2 - 2x\cos\left(\theta + \frac{4\pi}{n}\right) + 1\right\} \dots \left\{x^2 - 2x\cos\left(\theta + \frac{2n-2}{n}\pi\right) + 1\right\} \quad \dots(1)$$

By dividing by x^n we have

$$x^n + \frac{1}{x^n} - 2\cos n\theta = \left\{x + \frac{1}{x} - 2\cos 2\theta\right\}\left\{x + \frac{1}{x} - 2\cos\left(\theta + \frac{2\pi}{n}\right)\right\}$$

$$\dots\left\{x + \frac{1}{x} - 2\cos\left(\theta + \frac{2n-2}{n}\pi\right)\right\} \quad \dots(2)$$

The relation (2) may be written

$$x^n + \frac{1}{x^n} - 2\cos n\theta = \prod_{r=0}^{r=n-1}\left\{x + \frac{1}{2} - 2\cos\left(\theta + \frac{2r\pi}{n}\right)\right\}$$

where $\displaystyle\prod_{r=0}^{r=n-1}$ stands for the product for all integral values of r from $r = 0$ to $r = n - 1$ of the expression following it.

Similarly we may show that

$$x^{2n} - 2a^n x^n \cos n\theta + a^{2n}$$

$$= \{x^2 - 2ax\cos\theta + a^2\}\left\{x^2 - 2ax\cos\left(\theta + \frac{2\pi}{n}\right) + a^2\right\}$$

$$\left\{x^2 - 2ax\cos\left(\theta + \frac{4\pi}{n}\right) + a^2\right\}\dots\left\{x^2 - 2ax\cos\left(\theta + \frac{2n-2}{n}\pi\right) + a^2\right\} \quad \dots(3)$$

■ **116.** The proposition of the last article may also be proved by induction.

We shall first show that $x^n + \dfrac{1}{x^n} - 2\cos n\alpha$ is divisible by

$$x + \frac{1}{x} - 2\cos\alpha.$$

Let $x^n + \dfrac{1}{x^n} - 2\cos n\alpha$ be denoted by $\phi(n)$, and $x + \dfrac{1}{x} - 2\cos\alpha$ by λ, so that we have to show that $\phi(n)$ is divisible by λ, for all positive integral values of n.

Assume that this is true for $\phi(n - 1)$ and $\phi(n - 2)$.

We have then, by ordinary multiplication,

$$\left(x + \frac{1}{x}\right) \times \phi(n-1) = \left\{x + \frac{1}{x}\right\}\left\{x^{n-1} + \frac{1}{x^{n-1}} - 2\cos(n-1)\alpha\right\}$$

$$= \left(x^n + \frac{1}{x^n}\right) + \left(x^{n-2} + \frac{1}{x^{n-2}}\right) - 2\cos(n-1)\alpha \times \left(x + \frac{1}{x}\right)$$

$$= \left\{x^n + \frac{1}{x^n} - 2\cos n\alpha\right\}$$

$$+ \left\{x^{n-2} + \frac{1}{x^{n-2}} - 2\cos(n-2)\alpha\right\} - 2\cos(n-1)\alpha\left\{x + \frac{1}{x} - 2\cos\alpha\right\},$$

since $\quad 2\cos n\alpha + 2\cos(n-2)\alpha = 4\cos\alpha\cos(n-1)\alpha.$

Hence $\quad \left(x + \dfrac{1}{x}\right) \times \phi(n-1) = \phi(n) + \phi(n-2) - 2\lambda\cos(n-1)\alpha.$

∴ $\quad \phi(n) = \left(x + \dfrac{1}{x}\right)\phi(n-1) - \phi(n-2) + 2\lambda\cos(n-1)\alpha$ \qquad ...(1)

Now $\qquad \phi(1) = x + \dfrac{1}{x} - 2\cos\alpha = \lambda,$

and $\quad \phi(2) = x^2 + \dfrac{1}{x^2} - 2\cos 2\alpha = \left(x + \dfrac{1}{x} - 2\cos\alpha\right)\left(x + \dfrac{1}{x} + 2\cos\alpha\right)$

$$= \lambda\left(x + \frac{1}{x} + 2\cos\alpha\right),$$

so that $\phi(1)$ and $\phi(2)$ are divisible by λ.

Hence, putting $n = 3$ in (1), we see that $\phi(3)$ is divisible by λ.

Similarly putting, in (1), $n = 4, 5, 6, ...$ in succession we see that, by induction, $\phi(n)$ is divisible by λ for all values of n.

∴ $\qquad x^n + \dfrac{1}{x^n} - 2\cos n\alpha$ is divisible by $x + \dfrac{1}{x} - 2\cos\alpha.$

Again $x^n + \dfrac{1}{x^n} - 2\cos n\alpha = x^n + \dfrac{1}{x^n} - 2\cos n\left(\alpha + \dfrac{2\pi}{n}\right)$,

and is similarly divisible by

$$x + \dfrac{1}{x} - 2\cos\left[\alpha + \dfrac{2\pi}{n}\right].$$

Proceeding in this way we can show that it is divisible by

$$x + \dfrac{1}{x} - 2\cos\left[\alpha + \dfrac{4\pi}{n}\right],\ \ldots\ldots\ x + \dfrac{1}{x} - 2\cos\left[\alpha + \dfrac{n-1}{n}2\pi\right],$$

and hence obtain equation (2) of Art. 115.

■ **117. De Moivre's Property of the Circle.**

A geometrical meaning may be given to the equation (3) of Art. 115.

Let $ABCD$ ... be the angular points of a polygon of n sides which is inscribed in a circle of radius a, so that, O being the centre, we have

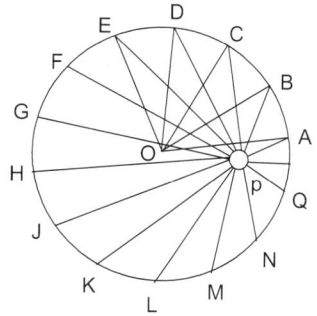

$$\angle AOB = \angle BOC = \angle COD = \ldots = \dfrac{2\pi}{n}.$$

Let P be a point within, or without the circle such that

$$OP = x, \text{ and } \angle POA = \theta.$$

Then

$$\angle POB = \theta + \dfrac{2\pi}{n},\ \angle POC = \theta + \dfrac{4\pi}{n},\ \ldots$$

and we have

$$PA^2 = OP^2 + OA^2 - 2OP \cdot OA \cos POA$$
$$= x^2 - 2ax \cos\theta + a^2,$$
$$PB^2 = OP^2 + OB^2 - 2OP \cdot OB \cos POB$$

$$= x^2 - 2ax\cos\left(\theta + \dfrac{2\pi}{n}\right) + a^2,$$

$$PC^2 = x^2 - 2ax\cos\left(\theta + \dfrac{4\pi}{n}\right) + a^2,$$

$$\cdot\ \cdot\ \cdot\ \cdot\ \cdot\ \cdot\ \cdot\ \cdot\ \cdot\ \cdot\ \cdot\ \cdot\ \cdot$$

Hence $PA^2 \cdot PB^2 \cdot PC^2 \ldots$ to n factors

$$\left\{x^2 - 2ax\cos\theta + a^2\right\}\left\{x^2 - 2ax\cos\left(\theta + \dfrac{2\pi}{n}\right) + a^2\right\}$$

$$\left\{x^2 - 2ax\cos\left(\theta + \dfrac{4\pi}{n}\right) + a^2\right\}\ldots \text{ to } n \text{ factors}$$

$$= x^{2n} - 2a^n x^n \cos n\theta + a^{2n}.$$

■ 118. Cotes' Property of the Circle.

In the preceding article let the point P lie on OA. *i.e.*, let it be on the line joining the centre to one of the angular points of the polygon.

In this case $\theta = 0$, and we have $PA^2 \cdot PB^2 \cdot PC^2$ ... to n factors

$$= x^{2n} - 2a^n x^n + a^{2n}$$
$$= (x^n - a^n)^2.$$

∴ $PA \cdot PB \cdot PC$ ... to n factors

$$= x^n - a^n \text{ or else } a^n - x^n.$$

The first of these values must be taken when P is outside the circle, on OA produced, so that $x > a$.

The second must be taken when P is within the circle.

We therefore have

$PA \cdot PB \cdot PC \cdot PD$ ... to n factors $= x^n \sim a^n$ ...(1)

Again let α, β, γ, δ ... be the middle points of the arcs AB, BC, CD, ... so that $A\alpha B\beta C\gamma$ ... is a polygon of $2n$ sides inscribed in the circle.

By (1) we have

$PA \cdot P\alpha \cdot PB \cdot P\beta \cdot PC \cdot P\gamma$ ... to $2n$ factors $= x^{2n} \sim a^{2n}$ ...(2)

Dividing (1) by (2), we get

$P\alpha \cdot P\beta \cdot P\gamma$ ... to n factors $= x^n + a^n$ ...(3)

The equation (3) may also be deduced directly from equation (3) of Art. 115 by putting $\theta = \dfrac{\pi}{n}$. We then have

$$\left(x^2 - 2ax\cos\frac{\pi}{n} + a^2\right)\left(x^2 - 2ax\cos\frac{3\pi}{n} + a^2\right)\left(x^2 - 2ax\cos\frac{5\pi}{n} + a^2\right)$$

...... to n factors $= x^{2n} - 2a^n x^n \cos \pi + a^{2n}$

$$= x^{2n} + 2a^n x^n + a^{2n} = (x^n + a^n)^2,$$

i.e. $P\alpha^2 \cdot P\beta^2 \cdot P\gamma^2$... to n factors $= (x^n + a^n)^2.$

This is relation (3).

■ 119. *To resolve into factors the expression* $x^n - 1$.

We have first to solve the equation

$$x^n - 1 = 0,$$

i.e., $x^n = 1 = \cos 2r\pi \pm i \sin 2r\pi,$

where r is any integer,

so that $x = [\cos 2r\pi \pm i\sin 2r\pi]^{1/n}$ ...(1)

First, let n be **even.**

As in Art. 24 the values of the expression (1) are

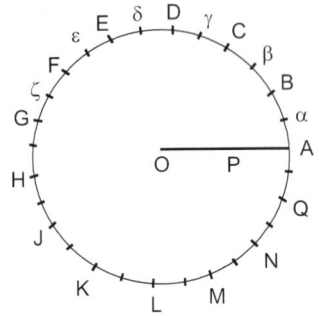

$$\cos 0 \pm i \sin 0, \ \cos \frac{2\pi}{n} \pm i \sin \frac{2\pi}{n}, \ \cos \frac{4\pi}{n} \pm i \sin \frac{4\pi}{n},$$

$$\ldots \cos \frac{n-2}{n}\pi \pm i \sin \frac{n-2}{n}\pi, \ \cos \frac{n\pi}{n} \pm i \sin \frac{n\pi}{n}.$$

But $\qquad \cos 0° \pm i \sin 0° = 1,$

and $\qquad \cos \dfrac{n\pi}{n} \pm i \sin \dfrac{n\pi}{n} = -1.$

Hence in this case the roots are the n quantities

$$\pm 1, \ \cos \frac{2\pi}{n} \pm i \sin \frac{2\pi}{n}, \ \cos \frac{4\pi}{n} \pm i \sin \frac{4\pi}{n}$$

$$\ldots \cos \frac{n-2}{n}\pi \pm i \sin \frac{n-2}{n}\pi.$$

The factors corresponding to the first of these pairs are $x - 1$ and $x + 1$, *i.e.*, the quadratic factor $x^2 - 1$.

Those corresponding to the second pair are

$$x - \cos \frac{2\pi}{n} - i \sin \frac{2\pi}{n} \text{ and } x - \cos \frac{2\pi}{n} + i \sin \frac{2\pi}{n},$$

i.e., the quadratic factor

$$x^2 - 2x \cos \frac{2\pi}{n} + 1.$$

Hence we get $\dfrac{n}{2}$ pairs of quadratic factors.

When multiplied together they give the correct coefficient for x^n, so that no numerical quantity need be prefixed to their product.

Hence, finally, when n is even,

$$x^n - 1 = \left(x^2 - 1\right)\left(x^2 - 2x\cos\frac{2\pi}{n} + 1\right)\left(x^2 - 2x\cos\frac{4\pi}{n} + 1\right)$$

$$\ldots\left(x^2 - 2x\cos\frac{n-2}{n}\pi + 1\right). \ \ldots \ldots \ldots \ldots \qquad (2)$$

Secondly, let n be **odd**.

As in Art. 24 the values of the expression (1) are now

$$\cos 0 \pm i \sin 0, \ \cos \frac{2\pi}{n} \pm i \sin \frac{2\pi}{n}, \ \cos \frac{4\pi}{n} \pm i \sin \frac{4\pi}{n}, \ \ldots$$

$$\ldots \cos \frac{n-3}{n}\pi \pm i \sin \frac{\pi-3}{n}\pi, \ \cos \frac{n-1}{n}\pi \pm i \sin \frac{n-1}{n}\pi.$$

The first pair reduces to the single root $+1$.

Taking the other pairs together, as before, we obtain, when n is odd,

$$x^n - 1 = (x - 1)\left\{x^2 - 2x\cos\frac{2\pi}{n} + 1\right\}\left\{x^2 - 2x\cos\frac{4\pi}{n} + 1\right\}\dots$$

$$\cdot\left\{x^2 - 2x\cos\frac{n-1}{n}\pi + 1\right\} \qquad \dots(3)$$

Hence we have

$$x^n - 1 = (x^2 - 1)\prod_{r=1}^{r=\frac{n}{2}-1}\left(x^2 - 2x\cos\frac{2r\pi}{n} + 1\right),$$

when n is even, and

$$x^n - 1 = (x - 1)\prod_{r=1}^{r=\frac{n-1}{2}}\left(x^2 - 2x\cos\frac{2r\pi}{n} + 1\right),$$

when n is odd.

These formulae can also be deduced from the fundamental one of Art. 115 by putting $n\theta = 2\pi$.

■ **120.** *To resolve $x^n + 1$ into factors.*

We must solve the equation

$$x^n + 1 = 0,$$

i.e., $\quad x^n = -1 = \cos(2r\pi + \pi) \pm i \sin(2r\pi + \pi),$

where r is any integer,

so that $\quad x = \{\cos(2r\pi + \pi) \pm i\sin(2r\pi + \pi)\}^{1/n}$

$$= \cos\frac{2r\pi + \pi}{n} \pm i\sin\frac{2r\pi + \pi}{n} \qquad \dots(1)$$

First, let n be **even.**

As in Art. 24, the values of the expression (1) are

$$\cos\frac{\pi}{n} \pm i\sin\frac{\pi}{n},\ \cos\frac{3\pi}{n} \pm i\sin\frac{3\pi}{n},\ \cos\frac{5\pi}{n} \pm i\sin\frac{5\pi}{n},$$

$$\dots\cos\frac{(n-1)\pi}{n} \pm i\sin\frac{(n-1)\pi}{n}.$$

The factors corresponding to the first of these pairs are

$$x - \cos\frac{\pi}{n} - i\sin\frac{\pi}{n} \text{ and } x - \cos\frac{\pi}{n} + i\sin\frac{\pi}{n},$$

i.e., the quadratic factor

$$x^2 - 2x\cos\frac{\pi}{n} + 1.$$

The quadratic factor corresponding to the second pair is

$$x^2 - 2x\cos\frac{3\pi}{n} + 1,$$

and so on.

Hence, as in the last article, when n is even, we have

$$x^n + 1 = \left(x^2 - 2x\cos\frac{\pi}{n} + 1\right)\left(x^2 - 2x\cos\frac{3\pi}{n} + 1\right)\dots$$

$$\dots\left[x^2 - 2x\cos\frac{(n-1)\pi}{n} + 1\right].$$

Secondly, let n be **odd.**

The values of the expression (1) are in this case

$$\cos\frac{\pi}{n} \pm i\sin\frac{\pi}{n}, \ \cos\frac{3\pi}{n} \pm i\sin\frac{3\pi}{n}, \ \dots$$

$$\cos\frac{(n-2)\pi}{n} \pm i\sin\frac{(n-2)\pi}{n}, \ \cos\frac{n\pi}{n} \pm i\sin\frac{n\pi}{n}.$$

The last pair of roots reduces to the single root -1, so that $x + 1$ is one of the required factors.

The quadratic factors corresponding to the successive pairs of roots are

$$x^2 - 2x\cos\frac{\pi}{n} + 1, \ x^2 - 2x\cos\frac{3\pi}{n} + 1, \ \dots$$

$$x^2 - 2x\cos\frac{n-2}{n}\pi + 1.$$

Hence, finally, when n is odd, we have

$$x^n + 1 = (x+1)\left(x^2 - 2x\cos\frac{\pi}{n} + 1\right)\left(x^2 - 2x\cos\frac{3\pi}{n} + 1\right)\dots$$

$$\dots\left[x^2 - 2x\cos\frac{(n-2)\pi}{n} + 1\right].$$

We have then

$$x^n + 1 = \prod_{r=0}^{r=\frac{n-2}{2}} \left(x^2 - 2x\cos\frac{2r+1}{n}\pi + 1\right),$$

when n is even, and

$$x^n + 1 = (x+1)\prod_{r=0}^{\frac{n-3}{2}} \left(x^2 - 2x\cos\frac{2r+1}{n}\pi + 1\right),$$

when n is odd.

These formulae can be deduced from the fundamental one of Art. 115 by putting $n\theta = \pi$.

▨ **121. EXAMPLE 1.** *Express as a product of n factors the quantities*

$$\cos n\phi - \cos n\theta \text{ and } \cosh n\phi - \cos n\theta.$$

In equation (2) of Art. 115 put $x = e^{\phi i}$, so that $x^{-1} = e^{-\phi i}$, and hence

$$x + x^{-1} = e^{\phi i} + e^{-\phi i} = 2 \cos \phi,$$

and

$$x^n + x^{-n} = e^{n\phi i} + e^{-n\phi i} = 2 \cos n\phi.$$

We then have

$$2 \cos n\phi - 2 \cos n\theta = (2 \cos \phi - 2 \cos \theta)\left[2\cos\phi - 2\cos\left(\theta + \frac{2\pi}{n}\right)\right]$$

$$\left[2\cos\phi - 2\cos\left(\theta + \frac{4\pi}{n}\right)\right]...... \text{ to } n \text{ factors,}$$

i.e.,

$$\cos n\phi - \cos n\theta = 2^{n-1}\{\cos\phi - \cos\theta\}\left\{\cos\phi - \cos\left(\theta + \frac{2\pi}{n}\right)\right\}...$$

$$...\left\{\cos\phi - \cos\left(\theta + \frac{2n-2}{n}\pi\right)\right\}$$

$$= 2^{n-1}\prod_{r=0}^{r=n-1}\left\{\cos\phi - \cos\left(\theta + \frac{2r\pi}{n}\right)\right\}.$$

Similarly by putting $x = e^{\phi}$ we have

$$\cosh n\phi - \cos n\theta$$

$$= 2^{n-1}[\cosh\phi - \cos\theta]\left[\cosh\phi - \cos\left(\theta + \frac{2\pi}{n}\right)\right]......$$

$$\left[\cosh\phi - \cos\left(\theta + \frac{2n-2}{n}\pi\right)\right].$$

EXAMPLE 2. *If n be even, prove that*

$$2^{\frac{n-1}{2}}\sin\frac{2\pi}{2n}\sin\frac{4\pi}{2n}\sin\frac{6\pi}{2n}.....\sin\frac{n-2}{2n}\pi = \sqrt{n}.$$

In equation (2) of Art. 119 put n equal to unity.

Then, since

$$\frac{x^{n-1}}{x^2 - 1} = \frac{x^{n-1} + x^{n-2} + ... + x + 1}{x + 1},$$

therefore, when x is unity,

$$\frac{x^n - 1}{x^2 - 1} = \frac{n}{2}.$$

Hence we have

$$\frac{n}{2} = \left(2 - 2\cos\frac{2\pi}{n}\right)\left(2 - 2\cos\frac{4\pi}{n}\right)......\left(2 - 2\cos\frac{n-2}{n}\pi\right),$$

i.e.,

$$n = 2 \cdot 4\sin^2\frac{2\pi}{2n} \cdot 4\sin^2\frac{4\pi}{2n}......4\sin^2\frac{n-2}{2n}\pi,$$

there being $\frac{n}{2} - 1$ factors,

$$= 2^{n-1} \cdot \sin^2 \frac{2\pi}{2n} \sin^2 \frac{4\pi}{2n} \ldots \ldots \sin^2 \frac{n-2}{2n}\pi.$$

Hence $$\pm\sqrt{n} = 2^{\frac{n-1}{2}} \sin \frac{2\pi}{2n} \sin \frac{4\pi}{2n} \ldots \ldots \sin \frac{n-2}{2n}\pi \qquad \ldots(1)$$

Each of the angles $\dfrac{2\pi}{2n}, \dfrac{4\pi}{2n}, \ldots, \dfrac{n-2}{n^2}\pi$ is less than a right angle, so that each of the sines on the right-hand side of (1) is positive.

On the left-hand side we therefore replace the ambiguity by the positive sign and have the required result.

EXAMPLES XX

Factorise the following quantities.

1. $x^6 + 2x^3 \cos 120° + 1$.   **2.** $x^8 + 2x^4 \cos 60° + 1$.

3. $x^{10} + 2x^5 \cos \dfrac{\pi}{3} + 1$.   **4.** $x^{12} + x^6 + 1$.

5. $x^{14} + x^7 + 1$.   **6.** $x^5 - 1$.   **7.** $x^6 + 1$.

8. $x^7 - 1$.   **9.** $x^9 + 1$.   **10.** $x^{10} - 1$.

11. $x^{13} + 1$.   **12.** $x^{14} - 1$.   **13.** $x^{20} + 1$.

14. If n be even, prove that

$$2^{\frac{n-1}{2}} \sin\frac{\pi}{2n}\sin\frac{3\pi}{2n}\sin\frac{5\pi}{2n}\ldots\ldots\sin\frac{n-1}{2n}\pi = 1$$

$$= 2^{\frac{n-1}{2}} \cos\frac{\pi}{2n}\cos\frac{3\pi}{2n}\ldots\ldots\cos\frac{n-1}{2n}\pi.$$

15. If n be odd, prove that

$$2^{\frac{n-1}{2}} \sin\frac{2\pi}{2n}\sin\frac{4\pi}{2n}\ldots\sin\frac{n-1}{2n}\pi = \sqrt{n} = 2^{\frac{n-1}{2}} \cos\frac{\pi}{2n}\cos\frac{3\pi}{2n}\ldots\cos\frac{n-1}{2n}\pi$$

and that

$$2^{\frac{n-1}{2}} \sin\frac{\pi}{2n}\sin\frac{3\pi}{2n}\ldots\sin\frac{n-2}{2n}\pi = 1 = 2^{\frac{n-1}{2}} \cos\frac{2\pi}{2n}\cos\frac{4\pi}{2n}\ldots\cos\frac{n-1}{2n}\pi.$$

16. Prove that $\sin\dfrac{\pi}{n}\sin\dfrac{2\pi}{n}\ldots\ldots\sin\dfrac{n-1}{n}\pi = \dfrac{n}{2^{n-1}}$.

17. If n be odd, prove that

$$\tan\frac{\pi}{n}\tan\frac{2\pi}{n}\tan\frac{3\pi}{n}\ldots\tan\frac{\frac{1}{2}(n-1)\pi}{n} = \sqrt{n}.$$

18. Show that $\cos n\theta$

$$= 2^{n-1}\left(\cos\theta - \cos\frac{\pi}{2n}\right)\left(\cos\theta - \cos\frac{3\pi}{2n}\right)\ldots\ldots\left(\cos\theta - \cos\frac{2n-1}{2n}\pi\right).$$

Prove that

19. $\sin n\phi = 2^{n-1} \sin\phi\sin\left(\phi + \dfrac{\pi}{n}\right)......\sin\left(\phi + \dfrac{n-1}{n}\pi\right)$

$= 2^{n-1} \prod\limits_{r=0}^{r=n-1} \sin\left(\phi + \dfrac{r\pi}{n}\right).$

[*Put x* = 1, *and* $\theta = 2\phi$, *in the equation of Art.* 115.]

20. $\cos n\phi = 2^{n-1} \sin\left(\phi + \dfrac{\pi}{2n}\right)\sin\left(\phi + \dfrac{3\pi}{2n}\right)....\sin\left[\phi + \dfrac{2n-1}{2n}\right].$

[*Change* ϕ *into* $\phi + \dfrac{\pi}{2n}$ *in the formula of the preceding question.*]

21. $2^{n-1} \cos\phi\cos\left(\phi + \dfrac{\pi}{n}\right)\cos\left(\phi + \dfrac{2\pi}{n}\right)......\cos\left(\phi + \dfrac{n-1}{n}\pi\right)$

$= (-1)^{n/2} \sin n\phi$, when *n* is even,

and $= (-1)^{n-1/2} \cos n\phi$, when *n* is odd.

[*Change* ϕ *into* $\phi + \dfrac{\pi}{2}$ *in the result of Ex* 19.]

T.P: (II) – 6

22. $2^{n-1} \cos\dfrac{\pi}{2n}\cos\dfrac{3\pi}{2n}\cos\dfrac{5\pi}{2n}......\cos\dfrac{2n-1}{2n}\pi = \cos\dfrac{n\pi}{2}.$

23. $2^{n-1} \sin\dfrac{\pi}{2n}\sin\dfrac{3\pi}{2n}\sin\dfrac{5\pi}{2n}......\sin\dfrac{2n-1}{2n}\pi = 1.$

24. $\cos\dfrac{\pi}{n}\cos\dfrac{2\pi}{n}......\cos\dfrac{(2n-1)\pi}{n} = \dfrac{(-1)^n - 1}{2}.$

25. Prove that

$$\dfrac{x^n - a^n\cos n\theta}{a^{2n} - 2a^n x^n \cos n\theta + a^{2n}} = \dfrac{1}{nx^{n-1}} \sum\limits_{r=0}^{r=n-1} \dfrac{x - a\cos\left(\theta + \dfrac{2r\pi}{n}\right)}{x^2 - 2ax\cos\left(\theta + \dfrac{2n\pi}{n}\right) + a^2}$$

[*In* (3) *of Art.* 115 *change x into x* + *h, expand and equate coefficients of h. Or take logarithms and differentiate with respect to x.*]

26. The circumference of a circle of radius *r* is divided into 2*n* equal parts at points P_1, P_2, ... P_{2n}; if chords be drawn from P_1 to the other points, prove that

$$P_1P_2 \cdot P_1P_2 ......P_1P_n = r^{n-1}\sqrt{n}.$$

Also, if *O* be the middle point of the arc P_1P_{2n}, prove that

$$OP_1 \cdot OP_2 ......OP_n = \sqrt{2}r^n.$$

27. If $A_1A_2 ... A_{2n+1}$ be a regular polygon of 2*n* + 1 sides, inscribed in a circle of radius *a*, and OA_{n+1} be a diameter, prove that

$$OA_1 \cdot OA_2 ...... OA_n = a^n.$$

28. $A_1A_2 \ldots A_n$ is a regular polygon of n sides. From O the centre of the polygon a line is drawn meeting the incircle in P_1 and the circumcircle in P_2.

Prove that the product of the perpendiculars on the sides drawn from P_1 is to the product of the perpendiculars from P_2 as

$$\cos^n \frac{\pi}{n} \cot^2 \frac{n\theta}{2} \text{ to } 1,$$

θ being the angle between OPP_1 and OA_1.

29. $ABCD \ldots$ is a regular polygon, of n sides, which is inscribed in a circle of radius a and centre O; prove that

$$PA^2 \cdot PB^2 \cdot PC^2 \ldots = r^{2n} - 2a^n r^n \cos n\theta + a^{2n},$$

where OP is r and the angle POA is θ.

Prove also that the sum of the angles that $AP, BP, CP, \ldots$ make, with $OA, OB, OC, \ldots$ produced is $\tan^{-1} \dfrac{r^n \sin n\theta}{r^n \cos n\theta - a^n}$.

Resolution of sin θ and cos θ into factors.

■ **122.** *To express sin θ as a product of an infinite series of factors.*

We have
$$\sin\theta = 2\sin\frac{\theta}{2}\cos\frac{\theta}{2}$$

$$= 2\sin\frac{\theta}{2}\sin\left(\frac{\pi}{2}+\frac{\theta}{2}\right) \qquad \ldots (1)$$

Similarly in (1) changing θ into $\dfrac{\theta}{2}$ and $\dfrac{\pi}{2}+\dfrac{\theta}{2}$ successively, we have

$$\sin\frac{\theta}{2} = 2\sin\frac{\theta}{2^2}\sin\left(\frac{\pi}{2}+\frac{\theta}{2^2}\right) = 2\sin\frac{\theta}{2^2}\sin\left(\frac{2\pi}{2^2}+\frac{\theta}{2^2}\right),$$

and
$$\sin\left(\frac{\pi}{2}+\frac{\theta}{2}\right) = 2\sin\left(\frac{\pi}{2^2}+\frac{\theta}{2^2}\right)\cdot\sin\left(\frac{\pi}{2}+\frac{\pi}{2^2}+\frac{\theta}{2^2}\right)$$

$$= 2\sin\left(\frac{\pi}{2^2}+\frac{\theta}{2^2}\right)\cdot\sin\left(\frac{3\pi}{2^2}+\frac{\theta}{2^2}\right).$$

Substituting these values in the right-hand side of (1) we have, after rearranging,

$$\sin\theta = 2^3 \sin\frac{\theta}{2^2}\sin\frac{\pi+\theta}{2^2}\sin\frac{2\pi+\theta}{2^2}\sin\frac{3\pi+\theta}{2^2} \qquad \ldots(2)$$

Applying once more the formula (1) to each of the terms on the right-hand of (2) and arranging, we have

$$\sin\theta = 2^7 \sin\frac{\theta}{2^3}\sin\frac{\pi+\theta}{2^3}\sin\frac{2\pi+\theta}{2^3}\sin\frac{3\pi+\theta}{2^3}\sin\frac{4\pi+\theta}{2^3}$$

$$\sin\frac{5\pi+\theta}{2^3}\sin\frac{6\pi+\theta}{2^3}\sin\frac{7\pi+\theta}{2^3} \qquad \ldots(3)$$

Continuing this process we have finally

$$\sin\theta = 2^{p-1}\sin\frac{\theta}{p}\sin\frac{\pi+\theta}{p}\sin\frac{2\pi+\theta}{p}\ldots\sin\frac{(p-1)\pi+\theta}{p} \qquad \ldots(4)$$

where p is a power of 2.

The last factor in (4)

$$= \sin\left[\pi - \frac{\pi-\theta}{p}\right] = \sin\frac{\pi-\theta}{p}.$$

The last factor but one

$$= \sin\frac{(p-2)\pi+\theta}{p} = \sin\left[\pi - \frac{2\pi-\theta}{p}\right] = \sin\frac{2\pi-\theta}{p}.$$

and so on.

Hence, taking together the second and last factors, the third and next to last, and so on, the equation (4) becomes

$$\sin\theta = 2p^{-1}\sin\frac{\theta}{p}\left\{\sin\frac{\pi+\theta}{p}\sin\frac{\pi-\theta}{p}\right\}\left\{\sin\frac{2\pi+\theta}{p}\sin\frac{2\pi-\theta}{p}\right\} \qquad \ldots(5)$$

The last factor is

$$\sin\frac{\frac{p}{2}\pi+\theta}{p}$$

which

$$= \sin\left(\frac{\pi}{2} + \frac{\theta}{p}\right) = \cos\frac{\theta}{p}.$$

Hence (5) is

$$\sin\theta = 2^{p-1}\sin\frac{\theta}{p}\left[\sin^2\frac{\pi}{p} - \sin^2\frac{\theta}{p}\right]\left[\sin^2\frac{2\pi}{p} - \sin^2\frac{\theta}{p}\right]\ldots$$

$$\ldots\left[\sin^2\frac{\left(\frac{p}{2}-1\right)\pi}{p} - \sin^2\frac{\theta}{p}\right]\cdot\cos\frac{\theta}{p} \qquad \ldots(6)$$

Divide both sides of (6) by $\sin\frac{\theta}{p}$ and make θ zero.

Since

$$\left[\frac{\sin\theta}{\sin\dfrac{\theta}{p}}\right]_{\theta=0} = \left[p\frac{\sin\theta}{\theta}\cdot\frac{\dfrac{\theta}{p}}{\sin\dfrac{\theta}{p}}\right]_{\theta=0} = p,$$

we have

$$p = 2^{p-1} \cdot \sin^2 \frac{\pi}{p} \cdot \sin^2 \frac{2\pi}{p} \sin^2 \frac{3\pi}{p} \ldots \sin^2 \frac{\left(\frac{p}{2}-1\right)\pi}{p} \qquad \ldots(7)$$

Dividing (6) by (7), we have

$$\sin\theta = p\sin\frac{\theta}{p}\left[1 - \frac{\sin^2\frac{\theta}{p}}{\sin^2\frac{\pi}{p}}\right]\left[1 - \frac{\sin^2\frac{\theta}{p}}{\sin^2\frac{2\pi}{p}}\right]\left[1 - \frac{\sin^2\frac{\theta}{p}}{\sin^2\frac{3\pi}{p}}\right]\ldots$$

$$\ldots\left[1 - \frac{\sin^2\frac{\theta}{p}}{\sin^2\left(\frac{p}{2}-1\right)\frac{\pi}{p}}\right]\cos\frac{\theta}{p} \qquad \ldots(8)$$

Now make p indefinitely great.

Since

$$\left[p\sin\frac{\theta}{p}\right]_{p=\infty} = \left[\frac{\sin\frac{\theta}{p}}{\frac{\theta}{p}}\cdot\theta\right]_{p=\infty} = \theta \text{ (Art. 228, Part I.),}$$

$$\left[\frac{\sin^2\frac{\theta}{p}}{\sin^2\frac{\pi}{p}}\right]_{p=\infty} = \left[\frac{\sin^2\frac{\theta}{p}\frac{\pi^2}{p^2}\frac{\theta^2}{\theta^2}}{\frac{\theta^2}{p^2}\sin^2\frac{\pi}{p}\frac{\pi^2}{\pi^2}}\right] = \frac{\theta^2}{\pi^2} \text{ (Art. 228, Part I.),}$$

and so on, we have

$$\sin\theta = \theta\left(1 - \frac{\theta^2}{\pi^2}\right)\left(1 - \frac{\theta^2}{2^2\pi^2}\right)\left(1 - \frac{\theta^2}{3^2\pi^2}\right)\ldots \text{ ad inf.}$$

This theorem may be written in the form

$$\sin\theta = \theta\prod_{r=1}^{r=\infty}\left(1 - \frac{\theta^2}{r^2\pi^2}\right)$$

▌ **123.** *To express* $\cos\theta$ *as a product of an infinite series of factors.*

In equation (4) of Art. 122 write for θ the quantity $\frac{\pi}{2}+\theta$, and the equation becomes

$$\cos\theta = 2^{p-1}\sin\frac{\pi+2\theta}{2p}\sin\frac{3\pi+2\theta}{2p}\sin\frac{5\pi+2\theta}{2p}\ldots\sin\frac{(2p-1)\pi+2\theta}{2p} \qquad \ldots(1)$$

The last factor

$$= \sin\left[\pi - \frac{\pi - 2\theta}{2p}\right] = \sin\frac{\pi - 2\theta}{2p},$$

the last but one

$$= \sin\left[\frac{(2p-3)\pi + 2\theta}{2p}\right] = \sin\frac{3\pi - 2\theta}{2p},$$

and so on.

Hence taking the factors in pairs, as before, we have

$$\cos\theta = 2^{p-1}\left[\sin\frac{\pi + 2\theta}{2p}\sin\frac{\pi - 2\theta}{2p}\right]\left[\sin\frac{3\pi + 2\theta}{2p}\sin\frac{3\pi - 2\theta}{2p}\right]\dots$$

$$= 2^{p-1}\left[\sin^2\frac{\pi}{2p} - \sin^2\frac{2\theta}{2p}\right]\left[\sin^2\frac{3\pi}{2p} - \sin^2\frac{2\theta}{2p}\right] \qquad \dots(2)$$

In (2) make θ zero and we have

$$1 = 2^{p-1}\cdot\sin^2\frac{\pi}{2p}\cdot\sin^2\frac{3\pi}{2p}\cdot\sin^2\frac{5\pi}{2p} \qquad \dots(3)$$

Dividing (2) by (3), we have

$$\cos\theta = \left[1 - \frac{\sin^2\frac{2\theta}{2p}}{\sin^2\frac{\pi}{2p}}\right]\left[1 - \frac{\sin^2\frac{2\theta}{2p}}{\sin^2\frac{3\pi}{2p}}\right]\left[1 - \frac{\sin^2\frac{2\theta}{2p}}{\sin^2\frac{5\pi}{2p}}\right]\dots$$

$$\dots\left[1 - \frac{\sin^2\frac{2\theta}{2p}}{\sin^2\frac{(p-1)\pi}{2p}}\right] \qquad \dots(4)$$

In (4) make p infinite; then, as in the last article, we have

$$\cos\theta = \left[1 - \frac{4\theta^2}{\pi^2}\right]\left[1 - \frac{4\theta^2}{3^2\pi^2}\right]\left[1 - \frac{4\theta^2}{5^2\pi^2}\right]\dots \text{ ad inf.}$$

This theorem may be written in the form

$$\cos\theta = \prod_{r=1}^{r=\infty}\left\{1 - \frac{4\theta^2}{(2r-1)^2\pi^2}\right\}.$$

Since $\cos\theta = \dfrac{\sin 2\theta}{2\sin\theta}$, the product of cos θ may be derived from the products for sin 2θ and sin θ.

■ **124.** The equation (4) of Art. 122 may, by mean of Art. 115, be shown to be true for all integral values of p. For we have

$$x^2p - 2xp \cos p\phi + 1$$

$$= \left\{x^2 - 2x\cos\phi + 1\right\}\left\{x^2 - 2x\cos\left(\phi + \frac{2\pi}{p}\right) + 1\right\}$$

$$\left\{x^2 - 2x\cos\left(\phi + \frac{4\pi}{p}\right) + 1\right\}\ldots \text{ to } p \text{ factors.}$$

Put $x = 1$, and we have

$$2(1 - \cos p\phi) = \{2 - 2\cos\phi\}\left\{2 - 2\cos\left(\phi + \frac{2\pi}{p}\right)\right\}\ldots\ldots \text{ to } p \text{ factors.}$$

i.e., $$4\sin^2\frac{p\phi}{2} = 4\sin^2\frac{\phi}{2}\cdot 4\sin^2\left(\frac{\phi}{2} + \frac{\pi}{p}\right)\cdot 4\sin^2\left(\frac{\phi}{2} + \frac{\pi}{p}\right)\ldots \text{ to } p \text{ factors.}$$

Put $\dfrac{p\phi}{2} = \theta$, and extract the square root of both sides. We have then

$$\pm\sin\theta = 2p^{-1}\sin\frac{\theta}{p}\cdot\sin\frac{\pi+\theta}{p}\cdot\sin\frac{2\pi+\theta}{p}\ldots\ldots\sin\frac{(p-1)\pi+\theta}{p} \qquad \ldots(1)$$

If θ lie between 0 and π all the factors on the right-hand side of (1) are positive and so also is $\sin\theta$. Hence the ambiguity should be replaced by the positive sign.

If θ lie between π and 2π, all the factors on the right-hand side are positive except the last, which is negative.

Hence the product is negative and so also is $\sin\theta$, so that in this case also the positive sign is to be taken.

Similarly, in any other case it may be shown that the positive sign must be taken, and we have, for all integral values of p,

$$\sin\theta = 2p^{-1}\sin\frac{\theta}{p}\cdot\sin\frac{\pi+\theta}{p}\sin\frac{2\pi+\theta}{p}\ldots\ldots\sin\frac{(p-1)\pi+\theta}{p}$$

▨ **125.** *sinh θ and cosh θ in products.*

By Art. 68 we have

$$\sinh\theta = -\imath\sin(\theta\imath) \text{ and } \cosh\theta = \cos(\theta\imath).$$

Also the series of Arts. 122 and 123, being formed on the Addition Theorem are, by Art. 64, true when for θ we read $\theta\imath$.

$\therefore$ $$\sinh\theta = -i\times\theta i\left(1 - \frac{\theta^2 i^2}{\pi^2}\right)\left(1 - \frac{\theta^2 i^2}{2^2\pi^2}\right)\left(1 - \frac{\theta^2 i^2}{3^2\pi^2}\right) \qquad \ldots(1)$$

$$= \theta\left(1 + \frac{\theta^2}{\pi^2}\right)\left(1 + \frac{\theta^2}{2^2\pi^2}\right)\left(1 + \frac{\theta^2}{3^2\pi^2}\right)\ldots\ldots \text{ ad inf.}$$

and $$\cosh\theta = \left(1 - \frac{4\theta^2 i^2}{\pi^2}\right)\left(1 - \frac{4\theta^2 i^2}{3^2\pi^2}\right)\left(1 - \frac{4\theta^2 i^2}{5^2\pi^2}\right)\ldots \text{ ad inf.}$$

$$= \left(1 + \frac{4\theta^2}{\pi^2}\right)\left(1 + \frac{4\theta^2}{3^2\pi^2}\right)\left(1 + \frac{4\theta^2}{5^2\pi^2}\right)\dots \text{ ad inf.} (2).$$

The products (1) and (2) are convergent. For we know (C. Smith's *Algebra*, Art. 337) that the infinite product II $(1 + u_n)$ is convergent if the series $\sum u_n$ be convergent.

In the case of (1), $\sum u_n$

$$= \frac{\theta^2}{\pi^2}\left(1 + \frac{1}{2^2} + \frac{1}{3^2} + \frac{1}{4^2} + \dots\right),$$

and the latter series is known to be convergent.

■ **126. Sums of Powers of the Reciprocals of all Natural Numbers.**

From the results of Arts. 122 and 123 we can deduce the sums of some interesting series.

From Arts. 122 and 33 we have

$$\left(1 - \frac{\theta^2}{\pi^2}\right)\left(1 - \frac{\theta^2}{2^2\pi^2}\right)\left(1 - \frac{\theta^2}{3^2\pi^2}\right)\dots\dots \text{ad inf.}$$

$$= \frac{\sin\theta}{\theta} = 1 - \frac{\theta^2}{\underline{3}} + \frac{\theta^4}{\underline{5}} + \dots\dots \text{ ad inf.}$$

Taking the logarithms of both sides, we have

$$\log\left(1 - \frac{\theta^2}{\pi^2}\right) + \log\left(1 - \frac{\theta^2}{2^2\pi^2}\right) + \log\left(1 - \frac{\theta^2}{3^2\pi^2}\right) + \dots.$$

$$= \log\left[1 - \frac{\theta^2}{6} + \frac{\theta^4}{120} - \dots\right] \qquad \dots (1)$$

Now, by Art. 8, we have

$$\log\left(1 - \frac{\theta^2}{\pi^2}\right) = -\left[\frac{\theta^2}{\pi^2} + \frac{1}{2}\frac{\theta^4}{\pi^4} + \frac{1}{3}\frac{\theta^6}{\pi^6} + \dots\right],$$

$$\log\left(1 - \frac{\theta^2}{2^2\pi^2}\right) = -\left[\frac{\theta^2}{2^2\pi^2} + \frac{1}{2}\frac{\theta^4}{2^4\pi^4} + \frac{1}{3}\frac{\theta^6}{2^6\pi^6} + \dots\right],$$

$$\dots\dots\dots\dots$$

so that (1) gives

$$-\frac{\theta^2}{\pi^2}\left[\frac{1}{1^2} + \frac{1}{2^2} + \frac{1}{3^2} + \dots\right] - \frac{1}{2}\frac{\theta^4}{\pi^4}\left[\frac{1}{1^4} + \frac{1}{2^4} + \frac{1}{3^4} + \dots\right]$$

$$-\frac{1}{3}\frac{\theta^6}{\pi^6}\left[\frac{1}{1^6} + \frac{1}{2^6} + \frac{1}{3^6} + \dots\right]\dots\dots$$

$$= \log\left[1 - \left(\frac{\theta^2}{6} - \frac{\theta^4}{120} + \dots\right)\right]$$

$$= -\left(\frac{\theta^2}{6} - \frac{\theta^4}{120} + ...\right) - \frac{1}{2}\left(\frac{\theta^2}{6} - \frac{\theta^4}{120} + ...\right)^2 - ...$$

$$= -\frac{\theta^2}{6} + \theta^4\left(\frac{1}{120} - \frac{1}{2} \cdot \frac{1}{36}\right) - ...$$

$$= -\frac{\theta^2}{6} - \frac{\theta^4}{180} - ...... \qquad ...(2)$$

Since equation (2) is true for all values of θ the coefficients of θ^2 on both sides must be the same, and similarly those of θ^4, and so on.

Hence we have

$$-\frac{1}{\pi^2}\left(\frac{1}{1^2} + \frac{1}{2^2} + \frac{1}{3^2} + ...\text{ad inf.}\right) = -\frac{1}{6},$$

$$-\frac{1}{2}\frac{1}{\pi^4}\left(\frac{1}{1^4} + \frac{1}{2^4} + \frac{1}{3^4} + ...\text{ad inf.}\right) = -\frac{1}{180},......$$

Hence
$$\frac{1}{1^2} + \frac{1}{2^2} + \frac{1}{3^2} + ...\text{ ad inf.} = \frac{\pi^2}{6} \qquad ...(3)$$

and
$$\frac{1}{1^4} + \frac{1}{2^4} + \frac{1}{3^4} + ...\text{ ad inf.} = \frac{\pi^4}{90} \qquad ...(4)$$

. . . .

■ **127.** By proceeding in a similar manner with the result of Art. 123, we have

$$\left(1 - \frac{4\theta^2}{\pi^2}\right)\left(1 - \frac{4\theta^2}{3^2\pi^2}\right)\left(1 - \frac{4\theta^2}{5^2\pi^2}\right)...$$

$$= \cos\theta = 1 - \frac{\theta^2}{\underline{2}} + \frac{\theta^4}{\underline{4}}...,$$

so that

$$\log\left(1 - \frac{4\theta^2}{\pi^2}\right) + \log\left(1 - \frac{4\theta^2}{3^2\pi^2}\right) + \log\left(1 - \frac{4\theta^2}{5^2\pi^2}\right) + ...$$

$$= \log\left[1 - \frac{\theta^2}{2} + \frac{\theta^4}{24} - ...\right].$$

Hence as before

$$\frac{-4\theta^2}{\pi^2}\left(\frac{1}{1^2} + \frac{1}{3^2} + \frac{1}{5^2} + ...\right) - \frac{1}{2}\frac{16\theta^4}{\pi^4}\left(\frac{1}{1^4} + \frac{1}{3^4} + \frac{1}{5^4} + ...\right) + ...$$

$$= \log\left[1 - \left(\frac{\theta^2}{2} - \frac{\theta^4}{24} + ...\right)\right]$$

$$= -\left(\frac{\theta^2}{2} - \frac{\theta^4}{24} + ...\right) - \frac{1}{2}\left(\frac{\theta^2}{2} - \frac{\theta^4}{24} + ...\right)^2 + ...$$

$$= -\frac{\theta^2}{2} + \frac{\theta^4}{24} + ... - \frac{1}{2}\left(\frac{\theta^4}{4} - ...\right) - = -\frac{\theta^2}{2} - \frac{\theta^4}{12} - ...$$

Hence, equating coefficients of θ^2 and θ^4, we have

$$-\frac{4}{\pi^2}\left(\frac{1}{1^2} + \frac{1}{3^2} + \frac{1}{5^2} + ...\right) = -\frac{1}{2},$$

$$-\frac{8}{\pi^4}\left(\frac{1}{1^4} + \frac{1}{3^4} + \frac{1}{5^4} + ...\right) = -\frac{1}{12},$$

. . . .

and hence $\qquad \dfrac{1}{1^2} + \dfrac{1}{3^2} + \dfrac{1}{5^2} + ... = \dfrac{\pi^2}{8}$ ...(1)

and $\qquad \dfrac{1}{1^4} + \dfrac{1}{3^4} + \dfrac{1}{5^4} + ... = \dfrac{\pi^4}{96}$ ...(2)

. . . .

■ **128. Wallis' Formula.**

In the expression of Art. 122 put $\theta = \dfrac{\pi}{2}$, and we have

$$1 = \frac{\pi}{2}\left[1 - \frac{1}{2^2}\right]\left[1 - \frac{1}{4^2}\right]\left[1 - \frac{1}{6^2}\right] ...... \text{ ad inf.}$$

$$= \frac{\pi}{2} \cdot \frac{1 \cdot 3}{2^2} \cdot \frac{3 \cdot 5}{4^2} \cdot \frac{5 \cdot 7}{6^2} ...... \frac{(2n-3)(2n-1)}{(2n-2)^2} \cdot \frac{(2n-1)(2n+1)}{(2n)^2},$$

where n is infinite,

i.e., $\qquad \dfrac{2}{\pi} = \dfrac{1^2 \cdot 3^2 \cdot 5^2 \cdot 7^3 ......(2n-1)^2 \cdot (2n+1)}{2^2 \cdot 4^2 \cdot 6^2 ......(2n)^2},$

i.e., $\qquad \dfrac{2 \cdot 4 \cdot 6......2n}{1 \cdot 3 \cdot 5......(2n-1)} = \sqrt{\dfrac{\pi}{2}(2n+1)}$, where n is infinite.

It follows that when n is very great (but not necessarily infinite) then

$$\frac{2 \cdot 4 \cdot 6...2n}{1 \cdot 3 \cdot 5......(2n-1)} = \sqrt{\frac{\pi}{2}(2n+1)} \text{ very nearly}$$

$$= \sqrt{n\pi}, \text{ ultimately.}$$

This is called Wallis' Formula, and gives in a simple form a very near approach to the product of the first n even numbers divided by the first n odd numbers when n is very great.

▦ **129. EXAMPLE:** *Prove that*

$$\tan\theta = 8\theta\left[\frac{1}{\pi^2 - 4\pi^2} + \frac{1}{3^2\pi^2 - 4\theta^2} + \frac{1}{5^2\pi^2 - 4\theta^2} + \ldots\ldots\right].$$

From Art. 123 we have

$$\log\cos\theta = \log\left(1 - \frac{4\theta^2}{\pi^2}\right) + \log\left(1 - \frac{4\theta^2}{3^2\pi^2}\right) + \log\left(1 - \frac{4\theta^2}{5^2\pi^2}\right) + \ldots\ldots \qquad \ldots(1)$$

In this equation substituting $\theta + h$ for θ we have

$$= \log\cos(\theta + h) = \log\left[1 - \frac{4}{\pi^2}(\theta + h)^2\right] + \log\left[1 - \frac{4}{3^2\pi^2}(\theta + h)^2\right] + \ldots\ldots \qquad \ldots(2)$$

Now $\log\cos(\theta + h) = \log[\cos\theta\,(\cos h - \tan\theta\sin h)]$

$$= \log\cos\theta + \log\left[1 - \frac{h^2}{\underline{2}} + \ldots\ldots - \tan\theta\left(h - \frac{h^2}{\underline{3}} + \ldots\ldots\right)\right] \qquad \text{(Art. 33)}$$

$= \log\cos\theta + \log[1 - h\tan\theta + \text{higher powers of } h]$

$= \log\cos\theta - h\tan\theta + \text{powers of } h.$ (Art. 8.)

Also $\log\left[1 - \dfrac{4}{\pi^2}(\theta + h)^2\right] = \log\dfrac{\pi^2 - 4\theta^2}{\pi^2} + \log\left[1 - \dfrac{8\theta h}{\pi^2 - 4\theta^2} + \ldots\ldots\right]$

$$= \log\left[1 - \frac{4\theta^2}{\pi^2}\right] - \frac{8\theta h}{\pi^2 - 4\theta^2} + \text{powers of } h,$$

and $\log\left[1 - \dfrac{4}{3^2\pi^2}(\theta + h)^2\right]$

$$= \log\left[1 - \frac{4\theta^2}{3^2\pi^2}\right] - \frac{8\theta h}{3^2\pi^2 - 4\theta^2} + \text{powers of } h.$$

. . . .

Substituting these values in (2) and equating on each side the coefficients of $-h$ we have

$$\tan\theta = \frac{8\theta}{\pi^2 - 4\theta^2} + \frac{8\theta}{3^2\pi^2 - 4\theta^2} + \frac{8\theta}{5^2\pi^2 - 4\theta^2} + \ldots \qquad \ldots(3)$$

$$= \sum_{r=0}^{r=\infty} \frac{8\theta}{(2r+1)^2\pi^2 - 4\theta^2}.$$

The series (3) may also be written

$$\tan\theta = \frac{2}{\pi - 2\theta} - \frac{2}{\pi + 2\theta} + \frac{2}{3\pi - 2\theta} - \frac{2}{3\pi + 2\theta} + \ldots\ldots$$

[The student who is acquainted with the Differential Calculus will observe that equation (3) is obtained by differentiating (1) with respect to θ.]

■ **130. EXAMPLE:** *Prove that*

$$\cosh 2a - \cos 2\theta$$

$$= 2\sin^2\theta \left[1 + \frac{a^2}{\theta^2}\right]\left[1 + \left(\frac{a}{\pi+\theta}\right)^2\right]$$

$$\left[1 + \left(\frac{a}{\pi-\theta}\right)^2\right]\left[1 + \left(\frac{\alpha}{2\pi+\theta}\right)^2\right]\left[1 + \left(\frac{\alpha}{2\pi-\theta}\right)^2\right] \dots \text{ ad inf.}$$

$$= 2\sin^2\theta \; \Pi \left[1 + \left(\frac{\alpha}{\theta+r\pi}\right)^2\right],$$

where r is zero or any positive or any negative integer.

We have

$$\cosh 2\alpha - \cos 2\theta = \cos 2\alpha i - \cos 2\theta = 2 \sin(\theta + \alpha i) \sin (\theta - \alpha i)$$

$$= 2(\theta + \alpha i)\left[1 - \frac{(\theta+\alpha i)^2}{\pi^2}\right]\left[1 - \frac{(\theta+\alpha i)^2}{2^2\pi^2}\right]\dots$$

$$\times (\theta - \alpha i)\left[1 - \frac{(\theta-\alpha i)^2}{\pi^2}\right]\left[1 - \frac{(\theta-\alpha i)^2}{2^2\pi^2}\right] \qquad \dots(1)$$

Now,
$$\left[1 - \frac{(\theta+\alpha i)^2}{\pi^2}\right]\left[1 - \frac{(\theta-\alpha i)^2}{\pi^2}\right]$$

$$= \left[\frac{(\pi+\theta+\alpha i)(\pi-\theta-\alpha i)}{\pi^2}\right]\left[\frac{(\pi+\theta-\alpha i)(\pi-\theta+\alpha i)}{\pi^2}\right]$$

$$= \frac{(\pi+\theta)^2 + \alpha^2}{\pi^2} \cdot \frac{(\pi-\theta)^2 + \alpha^2}{\pi^2}.$$

Hence (1) gives

$$\cosh 2\alpha - \cos 2\theta = 2(\theta^2 + \alpha^2)\left[\frac{(\pi+\theta)^2+\alpha^2}{\pi^2}\right]\left[\frac{(\pi-\theta)^2+\alpha^2}{\pi^2}\right]\left[\frac{(2\pi+\theta)^2+\alpha^2}{2^2\pi^2}\right]$$

$$\left[\frac{(2\pi-\theta)^2+\alpha^2}{2^2\pi^2}\right]\dots\dots \text{ ad inf.} \qquad \dots(2)$$

In (2) put α = 0 and we have

$$2\sin^2\theta = 2\theta^2 \cdot \frac{(\pi+\theta)^2}{\pi^2} \cdot \frac{(\pi-\theta)^2}{\pi^2} \cdot \frac{(2\pi+\theta^2)}{2^2\pi^2} \cdot \frac{(2\pi-\theta)^2}{2^2\pi^2} \dots\dots \text{ ad inf.} \qquad \dots(3)$$

Dividing (2) by (3) we have

$$\cosh 2\alpha - \cos 2\theta$$

$$= 2\sin^2\theta\left[1+\frac{\alpha^2}{\theta^2}\right]\left[1+\left(\frac{\alpha}{\pi-\theta}\right)^2\right]\left[1+\left(\frac{\alpha}{\pi+\theta}\right)^2\right]\left[1+\left(\frac{\alpha}{2\pi-\theta}\right)^2\right]$$

$$\left[1+\left(\frac{\alpha}{2\pi+\theta}\right)^2\right]..... \text{ ad inf.}$$

The factors of cosh $2\alpha + \cos 2\theta$ may now be obtained by changing θ into $\theta+\frac{\pi}{2}$ and they are found to be $2\cos^2\theta\,\Pi\left\{1+\left(\frac{\alpha}{\theta+r\pi}\right)^2\right\}$ where r is any odd integer, positive or negative.

EXAMPLES XXI

Prove that

1. $\dfrac{1}{1^2}-\dfrac{1}{2^2}+\dfrac{1}{3^2}-\dfrac{1}{4^2}+......+ \text{ ad inf.} = \dfrac{\pi^2}{12}$.

2. $\dfrac{1}{1^6}+\dfrac{1}{2^6}+\dfrac{1}{3^6}+...... \text{ ad inf.} = 6\dfrac{(2\pi)^6}{\lfloor 9}$.

3. $\dfrac{1}{1\cdot 2}+\dfrac{1}{2\cdot 4}+\dfrac{1}{3\cdot 6}+\dfrac{1}{4\cdot 8}+...... \text{ ad inf.} = \dfrac{\pi^2}{12}$.

4. $\dfrac{1}{3^4}+\dfrac{3}{5^4}+\dfrac{6}{7^4}+\dfrac{10}{9^4}+... \text{ ad inf.} = \dfrac{\pi^2}{64}\left(1-\dfrac{\pi^2}{12}\right)$.

5. Prove that the sum of products, taken two and two together, of the reciprocals of the squares of all odd numbers is $\dfrac{\pi^4}{384}$.

6. Prove that the sum of products, taken two and two together, of the reciprocals of the squares of all numbers is $\dfrac{\pi^4}{120}$.

Prove that

7. $\cot\theta = \dfrac{1}{\theta}-\dfrac{2\theta}{\pi^2-\theta^2}-\dfrac{2\theta}{2^2\pi^2-\theta^2}-......$

$$= \dfrac{1}{\theta}+\dfrac{1}{\theta-\pi}+\dfrac{1}{\theta+\pi}+\dfrac{1}{\theta-2\pi}+\dfrac{1}{\theta+2\pi}+......+ \text{ ad inf.}$$

8. $\operatorname{cosec}\theta = \dfrac{1}{\theta}-\dfrac{1}{\theta-\pi}-\dfrac{1}{\theta+\pi}+\dfrac{1}{\theta-2\pi}+\dfrac{1}{\theta+2\pi}-\dfrac{1}{\theta-3\pi}-\dfrac{1}{\theta+3\pi}+...$

$$= \dfrac{1}{\theta}+2\theta\sum_{n=1}^{n=\infty}\dfrac{(-1)^n}{(\theta^2-n^2\pi^2)},$$

and hence that

$$\frac{1+\theta\cosec\theta}{2\theta^2}=\frac{1}{\theta^2}-\frac{1}{\theta^2-\pi^2}+\frac{1}{\theta^2-2^2\pi^2}-\ldots\ldots \text{ ad inf.}$$

$$\left[\textit{Use the relation } \cosec\theta=\frac{1}{2}\left(\tan\frac{\theta}{2}+\cot\frac{\theta}{2}\right)\right].$$

9. $\dfrac{1}{4\pi}\sec\theta=\dfrac{1}{\pi^2-4\theta^2}-\dfrac{3}{3^2\pi^2-4\theta^2}+\dfrac{5}{5^2\pi^2-4\theta^2}-\ldots\ldots \text{ ad inf.}$

$$\left[\textit{Use the relation } 2\sec\theta=\tan\left(\frac{\pi}{4}+\frac{\theta}{2}\right)+\cot\left(\frac{\pi}{4}+\frac{\theta}{2}\right)\right].$$

10. $\dfrac{1}{4}\sec^2\theta=\dfrac{1}{(\pi-2\theta)^2}+\dfrac{1}{(\pi+2\theta)^2}+\dfrac{1}{(3\pi-2\theta)^2}+\dfrac{1}{(3\pi+2\theta)^2}+\ldots \text{ ad inf.}$

[*Apply the process of Art.* 129 *to the result obtained in that article.*]

11. $\cosec^2\theta=\dfrac{1}{\theta^2}+\dfrac{1}{(\theta-\pi)^2}+\dfrac{1}{(\theta+\pi)^2}+\dfrac{1}{(\theta-2\pi)^2}+\dfrac{1}{(\theta+2\pi)^2}+\ldots \text{ ad inf.}$

Prove that

12. $\dfrac{\sin(\alpha-\theta)}{\sin\alpha}=\left(1-\dfrac{\theta}{\alpha}\right)\left(1+\dfrac{\theta}{\pi-\alpha}\right)\left(1-\dfrac{\theta}{\pi+\alpha}\right)\left(1+\dfrac{\theta}{2\pi-\alpha}\right)\left(1-\dfrac{\theta}{2\pi+\alpha}\right)\ldots\ldots$

$=\Pi\left(1-\dfrac{\theta}{\alpha+r\pi}\right)$, where r is any positive or negative integer or zero.

13. $\dfrac{\sin(\alpha+\theta)}{\sin\alpha}=\Pi\left(1+\dfrac{\theta}{\alpha+r\pi}\right)$, where r is any positive or negative integer, including zero.

14. $\dfrac{\cos(\alpha+\theta)}{\cos\alpha}=\left(1+\dfrac{2\theta}{\pi+2\alpha}\right)\left(1-\dfrac{2\theta}{\pi-2\alpha}\right)\left(1+\dfrac{2\theta}{3\pi+2\alpha}\right)\left(1-\dfrac{2\theta}{3\pi-2\alpha}\right)\ldots$

$=\Pi\left[1+\dfrac{2\theta}{2\alpha+r\pi}\right]$, where r is any odd integer positive or negative.

15. $\dfrac{\cos(\alpha-\theta)}{\cos\alpha}=\Pi\left[1-\dfrac{2\theta}{2\alpha+r\pi}\right]$, where r is any odd integer, positive or negative.

16. $\dfrac{\cos\theta+\cos\alpha}{1+\cos\alpha}=\left[1-\dfrac{\theta^2}{(\pi+\alpha)^2}\right]\left[1-\dfrac{\theta^2}{(\pi-\alpha)^2}\right]\left[1-\dfrac{\theta^2}{(3\pi+\alpha)^2}\right]\left[1-\dfrac{\theta^2}{(3\pi+a)^2}\right]\ldots$

$$=\Pi\left[1-\dfrac{\theta^2}{(r\pi+a)^2}\right].$$

where r is any odd integer positive or negative.

[*Multiply together the results of Exs.* 14 *and* 15 *and then change* 2θ *and* 2α *into* θ and α.]

17. $\dfrac{\cos\theta-\cos\alpha}{1-\cos\alpha}=\left\{1-\dfrac{\theta^2}{a^2}\right\}\left\{1-\dfrac{\theta^2}{(2\pi+\alpha)^2}\right\}\left\{1-\dfrac{\theta^2}{(2\pi-\alpha)^2}\right\}\left\{1-\dfrac{\theta^2}{(4\pi+\alpha)^2}\right\}\ldots\ldots$

$$=\Pi\left[1-\dfrac{\theta^2}{(\alpha+r\pi)^2}\right].$$

where r is any even positive or negative integer, including zero.

Hence deduce the factors of $\cosh x - \cos \alpha$.

18. $\dfrac{\sin \alpha - \sin \theta}{\sin \alpha} = \left(1 - \dfrac{\theta}{\alpha}\right)\left(1 - \dfrac{\theta}{\pi - \alpha}\right)\left(1 + \dfrac{\theta}{\pi + \alpha}\right)\left(1 + \dfrac{\theta}{2\pi - \alpha}\right)\left(1 - \dfrac{\theta}{2\pi + \alpha}\right)\cdots$

19. $2 \cosh \theta + 2 \cos \alpha$

$$= 4\cos^2 \dfrac{\alpha}{2}\left[1 + \dfrac{\theta^2}{(\alpha + \pi)^2}\right]\left[1 + \dfrac{\theta^2}{(\alpha - \pi)^2}\right]\cdots$$

$$= 4\cos^2 \dfrac{\alpha}{2}\Pi\left[1 + \dfrac{\theta^2}{(\alpha + r\pi)^2}\right],$$

where r is any odd integer positive or negative.

20. Prove that

$$\sinh nu = n\sinh u \prod_{r=1}^{r=n-1}\left[1 + \dfrac{\sinh^2 \dfrac{u}{2}}{\sin^2 \dfrac{r\pi}{2n}}\right],$$

and deduce the expression for sinh u in the form of an infinite product of quadratic factors in u.

[*Start with the result, when θ is zero, of Ex. 1, Art. 121. In this result put ϕ equal to zero and divide.*]

21. Prove that the value of the infinite product

$$\left(1 + \dfrac{1}{1^2}\right)\left(1 + \dfrac{1}{2^2}\right)\left(1 + \dfrac{1}{3^2}\right)\cdots\cdots \text{ ad inf.}$$

is $\dfrac{1}{\pi}\sinh \pi.$

22. A semicircle is divided into m equal parts and a concentric and similarly situated semi-circle is divided into n equal parts. Every point of section of one semicircle is joined to every point of section of the other. Find the arithmetic mean of the squares of the joining lines and prove that when m and n are indefinitely increased the result is

$a^2 + b^2 - \dfrac{8ab}{\pi^2}$, where a and b are the radii of the semicircles.

23. The radii of an infinite series of concentric circles are $a, \dfrac{a}{2}, \dfrac{a}{3}, \cdots$. From a point at a distance $c\ (> a)$ from their common centre a tangent is drawn to each circle. Prove that

$$\sin \theta_1 \sin \theta_2 \sin \theta_3 \cdots\cdots = \sqrt{\dfrac{c}{\pi\alpha}\sin \dfrac{\pi\alpha}{c}},$$

where $\theta_1, \theta_2, \theta_3 \ldots$ are the angles that the tangents subtended at t common centre.

24. An infinite straight line is divided by an infinite number of points into portions each of length a. If any point P be taken so that y is its distance from the straight line and x is its distance measured along the straight line from one of the points of division, prove that the sum of the squares of the reciprocals of the distances of the point P from all the points of division is

$$\frac{\pi}{\alpha y} \cdot \frac{\sinh\dfrac{2\pi y}{a}}{\cosh\dfrac{2\pi y}{a} - \cos\dfrac{2\pi x}{a}}.$$

[*Use the result of Ex.* 7]

25. If a, b, c ... denote all the prime numbers 2, 3, 5 ... prove that

$$\left(1-\frac{1}{a^2}\right)\left(1-\frac{1}{b^2}\right)\left(1-\frac{1}{c^2}\right)\cdots = \frac{6}{\pi^2},$$

and
$$\left(1+\frac{1}{a^2}\right)\left(1+\frac{1}{b^2}\right)\left(1+\frac{1}{c^2}\right)\cdots = \frac{15}{\pi^2}.$$

26. Prove that

$$\prod_{m=1}^{m=\infty}\left[1-\frac{x}{m^2-c^2}\right]$$

$$= \frac{c}{\sqrt{c^2+x}} \cdot \frac{\sin\left\{\pi\sqrt{c^2+x}\right\}}{\sin \pi c}.$$

CHAPTER 10

Principle of Proportional Parts

■ **131.** In the present chapter we shall consider the Principle of Proportional Parts, the truth of which we assumed in Chapter 11, Part I.

We then assumed that if n be any number and $n + 1$ the next number, whose logarithms were given in our tables, and if h be any fraction, then, to 7 places of decimals, it is true that

$$\frac{\log(n+h) - \log n}{\log(n+1) - \log n} = h.$$

The truth of this statement we shall now consider.

■ **132. Common Logarithms.** We have, by Art. 12,

$$\log_{10}(n+h) - \log_{10} n = \log_{10}\frac{n+h}{n} = \mu \log_e\left(\frac{1+h}{n}\right)$$

where $\mu = 0.43429448...$

Hence, by Art. 8, we have

$$\log_{10}(n+h) - \log_{10} n = \frac{\mu h}{n} - \frac{\mu}{2}\frac{h^2}{n^2} + \frac{\mu}{3}\frac{h^3}{n^3} \qquad \text{...(1)}$$

Now in seven-figure logarithm tables n contains 5 digits, *i.e.*, n is not less than 10000. Hence, if h be less than unity, we have $\dfrac{\mu}{2}\dfrac{h^2}{n^2}$ less than

$$\frac{1}{2}(0.43429448...)\times\frac{1}{10^8},$$

i.e., less than $\dfrac{0.21714724...}{10^8}$, *i.e.* $< 0.0000000021....$.

Also $\dfrac{\mu}{3}\dfrac{h^3}{n^3}$ is less than one-ten thousandth part of this.

Hence in (1) the omission of all the terms on the right-hand side after the first will make no difference at least as far as the *seventh* place of decimals. To seven places we therefore have

$$\log_{10}(n+h) - \log_{10} n = \frac{\mu h}{n}.$$

So $$\log_{10}(n+1) - \log_{10} n = \frac{\mu \cdot 1}{n}.$$

Hence, by division,

$$\frac{\log_{10}(n+h) - \log_{10} n}{\log_{10}(n+1) - \log_{10} n} = h.$$

The principal assumed is therefore always true for the logarithms of numbers as given in seven-figure-tables.

■ **133.** We may enquire what is the smallest number in the tables to which we can safely apply the principle of proportional parts. We must find that value of n which

makes $\dfrac{\mu h^2}{2n^2} < \dfrac{1}{10}$, so that $n^2 > \dfrac{u}{2} \cdot 10 \cdot h^3$.

The greatest value of h being unity, we then have

$$\pi^2 > \frac{\mu}{2} \cdot 10^7, \ i.e. > 2171472.4......$$

∴ $$n > 1473.$$

The number 1473 is therefore the required least number.

■ **134. Natural Sines.** Suppose we have a table calculated for successive differences of angles, such that the number of radians in these successive differences is h.

[In the case of our ordinary tables h = number of radians in 1′

$$= \frac{\pi}{60 \times 180} = 0.000290888..., \ i.e., \ h < 0.0003].$$

Also let k be less than h. Then our principal was that

$$\frac{\sin(\theta + k) - \sin\theta}{\sin(\theta + h) - \sin\theta} = \frac{k}{h}.$$

We shall examine this assumption.

We have

$$\sin(\theta + k) - \sin\theta = \sin\theta\cos k + \cos\theta\sin k - \sin\theta$$

$$= \sin\theta\left[1 - \frac{k^2}{\lfloor 2} + \frac{k^4}{\lfloor 4} - ...\right] + \cos\theta\left[k - \frac{k^3}{\lfloor 3} + ...\right] - \sin\theta \qquad \text{(Arts. 32 and 33)}$$

$$= k\cos\theta - \frac{k^2}{\lfloor 2}\sin\theta - \frac{k^3}{\lfloor 3}\cos\theta...$$

The ratio of the third term to the first $= \dfrac{1}{6}k^2$ and this is always less than $\dfrac{1}{6}(0.0003)^2$, $i.e.$ always less than 0.00000002. The third and higher terms may therefore be safely neglected, and we have

$$\sin(\theta + k) - \sin\theta = k\cos\theta - \frac{k^2}{|2}\sin\theta \qquad \dots(1)$$

The numerical ratio of the second term to the first term

$$= \frac{1}{2}k\tan\theta. \qquad \dots(2)$$

This ratio is small, *except when θ is nearly equal to* $\frac{\pi}{2}$.

Hence, except when the angle is nearly a right angle, the second term in (1) may be neglected, and we have

$$\sin(\theta + k) - \sin\theta = k\cos\theta.$$

So, $\sin(\theta + h) - \sin\theta = h\cos\theta,$

and hence $$\frac{\sin(\theta + k) - \sin\theta}{\sin(\theta + h) - \sin\theta} = \frac{k}{h} \qquad \dots(3)$$

When θ is very nearly a right angle we cannot say that

$$\sin(\theta + k) - \sin\theta = k\cos\theta,$$

and hence in this case the relation (3) does not hold and the difference in the sine is not proportional to the difference in the angle. In this case then the differences are **irregular.** At the same time the differences are **insensible;** for, when θ is nearly $\frac{\pi}{2}$, $k\cos\theta$ is very small. In fact $k\cos\theta$ has nothing but ciphers as far as the seventh place of decimals, so as long as θ is within a few minutes of a right angle. Also

$$\frac{k^2}{|2}\sin\theta \text{ is always} < \frac{(0.0003)^2}{2}, \text{ i.e.} < 0.00000005\dots$$

Hence when the angle is nearly a right angle a comparatively small change in the sine will correspond to a comparatively large change in the angle; also at the same time these changes are irregular.

▨ **135. Natural Cosines.** Since the cosine of an angle is equal to the sine of its complement this case reduces to that of the sine. The principle is therefore true except when the angle is nearly zero, in which case the differences are insensible and irregular.

▨ **136. Natural Tangents.** With the same notation as before we have

$$\tan(\theta + k) - \tan\theta = \frac{\tan\theta + \tan k}{1 - \tan\theta\tan k} - \tan\theta = \frac{\tan k\sec^2\theta}{1 - \tan\theta\tan k}$$

$$= \tan k\sec^2\theta(1 + \tan\theta\tan k + \tan^2\theta\tan^2 k\dots)$$

$$= \sec^2\theta\left[k + \frac{k^3}{3} + \dots\right]\left[1 + \tan\theta\left(k + \frac{k^3}{3} + \dots\right) + \tan^2\theta\left(k^2 + \dots\right)\right] \qquad \text{(Art. 34)}$$

$$= k\sec^2\theta + k^2\frac{\sin\theta}{\cos^3\theta} + k^3\sec^2\theta\left[\frac{1}{3} + \tan^2\theta\right] + \dots\dots \qquad \dots(1)$$

The third and the higher terms may be omitted as before, except when θ is nearly a right angle.

Unless the quantity $k^2 \dfrac{\sin\theta}{\cos^3\theta}$ be large we shall then have

$$\tan(\theta + k) - \tan\theta = k\sec^2\theta \qquad\qquad ...(2)$$

and the rule is approximately true.

When θ is $> \dfrac{\pi}{4}$ the second term of the equation (1) is $> 2k^2$, so that taking the greatest value of k, $viz.$ about 0.0003, this would give a significant figure in the seventh place. The principle is therefore not true for angles greater than $\dfrac{\pi}{4}$, when the differences of the tabulated angles are $1'$.

■ **137. Natural Cotangents.** As in the last article it can be shown that the principle must not be relied upon for angles between 0 and 45°.

■ **138. Natural Secant.** We have $\sec(\theta + k) - \sec\theta$

$$= \frac{1}{\cos\theta\cos k - \sin\theta\sin k} - \frac{1}{\cos\theta}$$

$$= \sec\theta\left[\frac{1}{1 - k\tan\theta - \dfrac{1}{2}k^2 ...} - 1\right]$$

$$= \sec\theta\left[k\tan\theta + k^2\left(\frac{1}{2} + \tan^2\theta\right) + ...\right]$$

$$= k\sec\theta\tan\theta + k^2\sec\theta\left(\frac{1}{2} + \tan^2\theta\right) + ... \qquad\qquad ...(1)$$

The ratio of the second to the first term

$$= k\frac{\dfrac{1}{2} + \tan^2\theta}{\tan\theta} = k\left[\frac{1}{2}\cot\theta + \tan\theta\right].$$

This is small except when θ is nearly zero or $\dfrac{\pi}{2}$. Hence, except in these two cases, we have

$$\sec(\theta + k) - \sec\theta = k\tan\theta\sec\theta$$

and the rule is proved.

When θ is small the term $k\sec\theta\tan\theta$ is very small, so that the differences are insensible besides being irregular.

When θ is nearly $\dfrac{\pi}{2}$ this term is great, so that the differences are not insensible.

■ **139. Natural Cosecant.** Just as in the case of the secant it may be shown that the differences are insensible and irregular when θ is nearly 90°, and irregular when θ is nearly zero. Otherwise the principle holds.

▨ 140. Tabular Logarithmic Sine. We have

$$L_{10} \sin(\theta + k) - L_{10} \sin\theta = \log_{10}\frac{\sin(\theta + k)}{\sin\theta}$$

$$= \log_{10}[\cos k + \cot\theta\sin k] = \log_{10}\left[1 + k\cot\theta - \frac{k^2}{2}\cdots\right] \qquad \text{(Arts. 32 and 33)}$$

$$= \mu\left[k\cot\theta - \frac{k^2}{2} - \frac{1}{2}k^2\cot^2\theta + \ldots\right] \qquad \text{(Arts. 8 and 12)}$$

$$= \mu k\cot\theta - \frac{uk^2}{2}\operatorname{cosec}^2\theta\ldots$$

The numerical ratio of the second term to the first

$$= \frac{1}{2}k\cdot\frac{1}{\sin\theta\cos\theta} = \frac{k}{\sin 2\theta}.$$

This is small except when θ is near zero or a right angle.

Hence, with the exception of these two cases, we have

$$L\sin(\theta + k) - L\sin\theta = \mu\cot\theta + k,$$

so that the rule holds in general.

If θ be small the term $\mu k\cot\theta$ is large, so that the differences are large as well as irregular. We cannot therefore apply the principle to small angles in the case of tables constructed with difference of $1'$.

Even if the tables were constructed for differences of $10''$ we are not sure of being free from error in the 7th place of decimals unless θ be $> 5°$.

If θ be nearly $\dfrac{\pi}{2}$ the terms $\mu k\cot\theta$ and $\dfrac{\mu k^2}{2}\operatorname{cosec}^2\theta$ are both small, so that if the angle be nearly a right angle the differences are insensible as well as irregular.

▨ 141. Tabular Logarithmic Cosine.
The rule holds approximately, since the cosine is the complement of the sine, except when the angle is small, in which case the differences are insensible as well as irregular, and except when the angle is nearly a right angle, in which case the differences are large.

▨ 142. Tabular Logarithmic Tangent. Here

$$L\tan(\theta + k) - L\tan\theta = \log_{10}\frac{\tan(\theta + k)}{\tan\theta}$$

$$= \log_{10}\frac{1 + \cot\theta\tan k}{1 - \tan\theta\tan k} = \log_{10}\left[\frac{1 + k\cot\theta}{1 - k\tan\theta}\right]$$

$$= \log_{10}[(1 + k\cot\theta)(1 + k\tan\theta + k^2\tan^2\theta + \ldots)]$$

$$= \log_{10}\left[1 + \frac{k}{\sin\theta\cos\theta} + \frac{k^2}{\cos^2\theta} + \ldots\right]$$

$$= \mu \left[\frac{k}{\sin\theta\cos\theta} + \frac{k^2}{\cos^2\theta} - \frac{1}{2} \frac{k^2}{\sin^2\theta\cos^2\theta} + ... \right] \qquad \text{(Arts. 8 and 12)}$$

$$= \frac{\mu k}{\sin\theta\cos\theta} - 2\mu k^2 \frac{\cos 2\theta}{\sin^2 2\theta} + ...$$

The numerical ratio of the second term to the first $= k \cot 2\theta$. This is small except when θ is near zero or a right angle.

Hence, with the exception of these two cases, we have

$$L\tan(\theta + k) - L\tan\theta = \frac{2\mu}{\sin 2\theta} \cdot k,$$

so that the principle is in general true.

In each of the exceptional cases $\dfrac{k}{\sin 2\theta}$ is not small, so that the differences are then irregular but not insensible.

The same statements are true for the tabular logarithmic cotangent.

■ **143. Tabular Logarithmic Secant and Cosecant.** We have

$$L\sec(\theta + k) - L\sec\theta = L\cos\theta - L\cos(\theta + k)$$

and $\qquad L\operatorname{cosec}(\theta + k) - L\operatorname{cosec}\theta = L\sin\theta - L\sin(\theta + k).$

Hence the results for the L sin and L cos are also true for the L cosec and L sec.

CHAPTER 11

Errors of Observations

■ **144.** We have up to the present assumed that it is possible to observe any angles perfectly accurately. In practice this is by no means the case. Our observations are liable to two classes two of errors, one due to the instruments themselves, which are hardly ever in *perfect* adjustment, and the other class due to mistakes on the part of the observer.

■ **145.** An error in any of our observations will clearly, in general, cause an error in the value of any quantity calculated from that observation. For example, if in Art. 192, Part I., there be a small error in the value of a, there will be a consequent error in the value of x which, as we see from the result of that article, depends on a.

■ **146.** The importance of an error in a length depends, in general, upon its ratio to that length. For example in measuring a piece of wood, about two metres long, a mistake of three centimetres would be a very serious error; in measuring a 500 metre racecourse a mistake of three centimetres would be not worth considering; whilst in measuring the distance from the Earth to the Moon an error of three centimetres would be absolutely inappreciable.

■ **147.** We shall assume that the errors we have to consider are so small that their squares (when measured in radians if they be angles) may be neglected and we shall give some examples of finding the errors in derived quantities.

We shall assume that our tables and calculations are correct, so that we have not to deal with mistakes in calculation but only with errors in the original observation.

▌ **148. EXAMPLE 1.** *MP (Fig Art. 42, Part I) is a vertical pole: at a point O distant a from its foot its angular elevation is found to be θ and its height then calculated; if there be an error δ in the observation of θ, find the consequent error in the height.*

The calculated height $h = a \tan \theta$, clearly.

If the error δ be in excess, the real elevation is $\theta - \delta$, and hence the real height $h' = a \tan (\theta - \delta)$.

Hence the error $h - h' = a \tan \theta - a \tan (\theta - \delta)$

$$= a \frac{\sin \delta}{\cos \theta \cos (\theta - \delta)} = a \sec^2 \theta \cdot \delta,$$

if we neglect squares and higher powers of δ.

The ratio of the error to the calculated height

135

$$= \delta \sec^2 \theta \div \tan \theta = \frac{2\delta}{\sin 2\theta}.$$

Except when $\sin 2\theta$ is small this ratio is small since δ is small. It is least when $\sin 2\theta$ is greatest, *i.e.*, when θ is $\dfrac{\pi}{4}$.

The ratio is large when θ is near zero and when it is near $\dfrac{\pi}{2}$.

Hence a *small* mistake in the angle makes a relatively large mistake in the calculated result when the angle subtended is very small or when it is very nearly $\dfrac{\pi}{2}$.

When θ is small, both the calculated height and the absolute error, *viz.* $a \tan \theta$ and $a \sec^2 \theta \cdot \delta$, are small, but the latter is great *compared with the former*.

When θ is nearly $90°$, both these quantities are great.

■ **EXAMPLE 2.** *The height of a tower is found as in Art. 192, Part I.; if there be an error θ in excess in the angle α, find the corresponding correction to be made in the height.*

The real value of α is $\alpha - \theta$; hence the real value of the height is found by substituting $\alpha - \theta$ for α in the obtained answer, and therefore

$$= a \frac{\sin(\alpha - \theta)\sin\beta}{\sin(\beta - \alpha + \theta)} = a \sin\beta \frac{\sin\alpha\cos\theta - \cos\alpha\sin\theta}{\sin(\beta - \alpha)\cos\theta + \cos(\beta - \alpha)\sin\theta}$$

$$= \frac{a\sin\alpha\sin\beta}{\sin(\beta - \alpha)} \cdot \frac{1 - \theta\cot\alpha}{1 + \theta\cot(\beta - \alpha)} \qquad \text{(Arts. 32 and 33)}$$

$$= \frac{a\sin\alpha\sin\beta}{\sin(\beta - \alpha)}[1 - \theta\cot\alpha][1 - \theta\cot(\beta - \alpha) + \ldots\ldots]$$

$$= \frac{a\sin\alpha\sin\beta}{\sin(\beta - \alpha)}[1 - \theta\{\cot(\beta - \alpha) + \cot\alpha\}]$$

$$= \frac{a\sin\alpha\sin\beta}{\sin(\beta - \alpha)} - \theta\frac{a\sin^2\beta}{\sin^2(\beta - \alpha)}.$$

The error in the calculated height is therefore $\theta \cdot \dfrac{a\sin^2\beta}{\sin^2(\beta - \alpha)}$, and is one of excess.

Also the ratio of the error to the calculated height

$$= \frac{\theta\sin\beta}{\sin\alpha\sin(\beta - \alpha)}.$$

EXAMPLE 3. *The angles of a triangle are calculated from the sides $a = 2$, $b = 3$ and $c = 4$, but it is found that the side c is overestimated by a small quantity δ; find the consequent errors in the angles.*

From the given values of the sides, we easily have

$$\cos A = \frac{7}{8}, \qquad \cos B = \frac{11}{16}, \qquad \cos C = -\frac{1}{4},$$

$$\sin A = \frac{2\sqrt{15}}{16}, \qquad \sin B = \frac{3\sqrt{15}}{16}, \quad \text{and } \sin C = \frac{4\sqrt{15}}{16}.$$

Corresponding to the value $4 - \delta$, let the values of the angles be $A - \theta_1$, $B - \theta_2$, and $C - \theta_3$.

Then
$$\cos(A - \theta_1) = \frac{3^2 + (4 - \delta)^2 - 2^2}{2(4 - \delta) \cdot 3} = \frac{21 - 8\delta}{24}\left(1 - \frac{\delta}{4}\right)^{-1},$$

i.e.
$$\cos A + \sin A \cdot \theta_1 = \frac{1}{24}[21 - 8\delta]\left[1 + \frac{\delta}{4}\right] = \frac{1}{24}\left[21 = \frac{11}{4}\delta\right], \quad \text{[Arts. 32 and 33]}$$

i.e.
$$\frac{7}{8} + \frac{2\sqrt{15}}{16}\theta_1 = \frac{7}{8} - \frac{11}{96}\delta,$$

so that
$$\theta_1 = -\frac{11\sqrt{15}}{180}\delta......\qquad\qquad ...(1)$$

Also
$$\cos(B - \theta_2) = \frac{(4 - \delta)^2 + 2^2 - 3^2}{2(4 - \delta) \cdot 2} = \frac{11 - 8\delta}{16}\left(1 - \frac{\delta}{4}\right)^{-1},$$

i.e.
$$\frac{11}{16} + \sin B \cdot \theta_2 = \frac{1}{16}[11 - 8\delta]\left[1 + \frac{\delta}{4}\right] = \frac{1}{16}\left[11 - \frac{21}{4}\delta\right],$$

i.e.
$$\frac{3\sqrt{15}}{16}\theta_2 = -\frac{21}{64}\delta,$$

so that
$$\theta_2 = -\frac{7\sqrt{15}}{60}\delta\qquad\qquad ...(2)$$

Also
$$\cos(C - \theta_2) = \frac{2^2 + 3^2 - (4 - \delta)^2}{2 \cdot 2 \cdot 3} = \frac{-3 + 8\delta}{12}.$$

i.e.
$$-\frac{1}{4} + \frac{4\sqrt{15}}{16}\theta_2 = -\frac{1}{4} + \frac{2\delta}{3},$$

so that
$$\theta_3 = \frac{8\sqrt{15}}{45}\delta.$$

The errors in the angles are therefore

$$\frac{-11\sqrt{15}}{180}\delta, \ \frac{-21\sqrt{15}}{180}\delta, \text{ and } \frac{32\sqrt{15}}{180}\delta \text{ radians.}$$

so that the smallest angle has the least error.

We note, as might have been assumed *a priori*, that the sum of the errors in the three angles is zero. This is necessarily so, since the sum of the angles of any triangle is always two right angles.

EXAMPLES XXII

1. The height of a hill is found by measuring the angles of elevation α and β of the top and bottom of a tower of height b on the top of the hill. Prove that the error in the height h caused by an error θ in the measurement of the angle α is $\theta \cdot \cos \beta \sec \alpha \operatorname{cosec} (\alpha - \beta)$ times the calculated height of the hill.

2. At a distance of 100 m from the foot of a tower the elevation of its top is found to be $30°$; find the greatest and least errors in its calculated height due to errors of $1'$ and 50 cm in the elevation and distance respectively.

3. In the example of Art. 196 (Part I.) find the errors in the calculated values of the flagstaff and tower due to an error δ in the observed value of α.

   If $a = 1000/m$, $\alpha = 30°$, $\beta = 15°$, and there be an error of 1 m in the value of α, calculate the numerical value of these errors.

4. AB is a vertical pole, and CD a horizontal line which when produced passes through B the foot of the pole. The tangents of the angles of elevation at C and D of the top of the pole are found to be $\frac{4}{3}$ and $\frac{3}{4}$ respectively. Find the height of the pole having given that $CD = 35$ metres.

   Prove that an error of in $1'$ the determination of the elevation at D will cause an error of approximately 8 cm in the calculated height of the pole.

5. The elevation of the summit of a tower is observed to be α at a station A and β at a station B, which is at a distance c from A in the direct horizontal line from the foot of the tower, and its height is thus found to be $\dfrac{c \sin \alpha \sin \beta}{\sin (\alpha - \beta)}$.

   If AB be measured not directly from the tower but horizontally and in a direction inclined at a small angle θ to the direct line, show that, to correct the height of the tower to the second order of small quantities, the quantity $\dfrac{c \cos \alpha \sin^2 \beta}{\cos \beta \sin (\alpha - \beta)} \dfrac{\theta^2}{2}$ must be subtracted.

6. A, B and C are three given points on a straight line; D is another point whose distance from B is found by observing that the angles ADB and CDB are equal and of an observed magnitude θ; prove that the error in the calculated length of DB consequent on a small error δ in the observed magnitude of θ, is

$$-\frac{2ab(a+b)^2 \sin\theta}{\left(a^2 + b^2 - 2ab\cos 2\theta\right)^{3/2}}\delta$$

   approximately, where $AB = a$ and $BC = b$.

7. In measuring the three sides of a triangle small errors x and y are made in two of them, a and b; prove that the error in the angle C will be $-\dfrac{y}{b}\cot A - \dfrac{x}{a}\cot B$, and find the errors in the other angles.

8. In a triangle ABC we have given that approximately $a = 36$ m, $b = 50$ m, and $C = \tan^{-1}\dfrac{3}{4}$; find what error in the given value of a will cause an error in the calculated value of c equal to that caused by an error of $5''$ in the measurement of C.

9. A triangle is solved from the parts $C = 15°$, $\alpha = \sqrt{6}$, and $b = 2$; prove that an error of $10''$ in the value of C would cause an error of about $13.66''$ in the calculated value of B.

10. Two sides b and c and the included angle A of a given triangle are supposed to be known; if there be a small error θ in the value of the angle A, prove that

    (1) the consequent error in the calculated value of B is
    $$-\theta \sin B \cos C \operatorname{cosec} A \text{ radians,}$$

    (2) the consequent error in the calculated value of a is $c \sin B . \theta$, and

    (3) the consequent error in the calculated area of the triangle is
    $$\theta \cot A \text{ times that area.}$$

11. There are errors in the sides a, b, and c of a triangle equal to x, y, and z respectively; prove that the consequent error in the calculated value of the circum-radius is
    $$\frac{1}{2}\cot A \cot B \cot C\left[x\sec A + y\sec B + z\sec C\right].$$

12. The area of a triangle is found by measuring the lengths of the sides and the limit of error possible, either in excess or defect, in measuring any length is n times that length, where n is small. Prove that in the case of the triangle whose sides are measured as 110, 81, and 59 metres, the limit to the error in the deduced area of the triangle is about $3.1433 \, n$ times that area.

13. The three sides of a triangle are measured and found to be nearly equal. If the measurements can be wrong one per cent in excess or defect, prove that the greatest error that can arise in calculating one of the angles is $80'$ nearly.

14. It is observed that the elevation of the summit of a mountain at each corner of a plane horizontal equilateral triangle is α; prove that the height of the mountain is
    $$\frac{1}{\sqrt{3}}a\tan\alpha,$$

    where a is the side of the triangle. If there be a small error n'' in the elevation at C, show that the true height is
    $$\frac{1}{\sqrt{3}}a\tan\alpha\left[1+\frac{\sin n''}{3\sin\alpha\cos\alpha}\right].$$

The student, who is acquainted with the Differential Calculus, will see that the results of some of the examples in this chapter may be more easily obtained by simple differentiation.

Thus, in Ex. 2. of p. 136, the height x of the tower
$$=\frac{a\sin\alpha\sin\beta}{\sin(\beta-\alpha)}.$$

If β be constant, and α vary, then, by differentiation,
$$\frac{dx}{d\alpha}=a\sin\beta\cdot\frac{\cos\alpha\sin(\beta-\alpha)+\sin\alpha\cos(\beta-\alpha)}{\sin^2(\beta-\alpha)},$$

$$\therefore \ \delta x=\frac{a\sin^2\beta}{\sin^2(\beta-\alpha)}\delta\alpha,$$

giving the small change δx in x due to small change $\delta\alpha$ in α.

Again, in Ex. 7 of p. 138, we have

$$\cos C = \frac{a^2 + b^2 - c^2}{2ab}.$$

Hence, c being constant, we have, on differentiation,

$$-\sin C \cdot \delta C = \frac{(2a\delta a + 2b\delta b) \cdot ab - (a^2 + b^2 - c^2)(a\delta b + b\delta a)}{2a^2 b^2}$$

$$= \frac{b \cdot 2ac \cos B \cdot \delta a + a \cdot 2bc \cos A \cdot \delta b}{2a^2 b^2}.$$

$$\therefore \quad \delta C = -\delta a \cdot \frac{c \cos B}{ab \sin C} - \delta b \cdot \frac{c \cos A}{ab \sin C} = -\frac{\delta a}{a} \cdot \cot B - \frac{\delta b}{b} \cdot \cot A.$$

Miscellaneous Propositions

Solution of a Cubic Equation.

■ **149.** The standard form of a cubic equation is
$$y^3 + 3ay^2 + 3by + c = 0.$$
Put $y = x - a$, and this equation becomes
$$x^3 - 3(a^2 - b)x + (2a^3 - 3ab + c) = 0,$$
i.e. it becomes of the form
$$x^3 - 3px + q = 0 \qquad \qquad ...(1)$$
Hence any cubic equation can be reduced to the form (1), which has no term containing x^2.

■ **150.** *To solve the equation $x^3 - 3px + q = 0$.*

Put $x = \dfrac{z}{n}$, and we have
$$z^3 - 3pn^2z + qn^3 = 0 \qquad \qquad ...(2)$$
Now, by Art. 107, we always have
$$\cos 3\theta = 4 \cos^3 \theta - 3 \cos \theta,$$
so that
$$\cos^3 \theta - \frac{3}{4} \cos \theta - \frac{1}{4} \cos 3\theta = 0 \qquad \qquad ...(3)$$

Now (2) and (3) are the same equation if
$$z = \cos \theta, \ 3pn^2 = \frac{3}{4}, \text{ and } -\frac{1}{4} \cos 3\theta = qn^3.$$

Hence
$$n = \left(\frac{1}{4p}\right)^{1/2},$$

and therefore
$$\cos 3\theta = -4q\left(\frac{1}{4p}\right)^{3/2} \qquad \qquad ...(4)$$

The equation (4) can always be solved (by means of the tables if necessary) if p be positive, and $4q\left(\dfrac{1}{4p}\right)^{3/2} < 1$,

i.e. if
$$q^2 < 4p^3.$$

[The student who is acquainted with the Theory of Equations will notice that is the case which cannot be solved by Cardan's Method. It is the case when the roots of the original cubic are all real.]

If θ be the smallest angle satisfying equation (4), then the values

$$\theta + \frac{2\pi}{3} \text{ and } \theta + \frac{4\pi}{3}$$

also satisfy it, so that the roots of the equation

$$x^3 - 3px + q = 0$$

are

$$\frac{1}{n}\cos\theta, \ \frac{1}{n}\cos\left(\theta + \frac{2\pi}{3}\right), \text{ and } \frac{1}{n}\cos\left(\theta + \frac{4\pi}{3}\right),$$

i.e.

$$2\sqrt{p}\cos\theta, \ 2\sqrt{p}\cos\left(\theta + \frac{2\pi}{3}\right), \text{ and } 2\sqrt{p}\cos\left(\theta + \frac{4\pi}{3}\right).$$

[See also Ex. 81, page 160]

■ **151. EXAMPLE.** *Solve the equation*

$$x^3 + 6x^2 + 9x + 3 = 0.$$

Put $\quad x = y - 2$, and the equation becomes

$$y^2 - 9y + 1 = 0$$

Put $y = \dfrac{z}{n}$, and the equation is

$$z^3 - 3n^3z + n^3 = 0 \qquad \qquad \text{...(1)}$$

Now, $\quad \cos^3\theta - \dfrac{3}{4}\cos\theta - \dfrac{1}{4}\cos 3\theta = 0 \qquad \qquad \text{...(2)}$

Equations (1) and (2) are the same if

$$z = \cos\theta, \ n^2 = \frac{1}{4}, \text{ and } -\frac{1}{4}\cos 3\theta = n^3,$$

i.e. if $\quad n = \dfrac{1}{2},$

and $\quad \cos 3\theta = -\dfrac{1}{2} = \cos 120° \qquad \qquad \text{...(3)}$

The roots of (3) are clearly

$$40°, \ 40° + 120°, \text{ and } 40° + 240°,$$

so that $\quad z = \cos 40°, \text{ or } \cos 160°, \text{ or } \cos 280°.$

∴ $\quad y = 2\cos 40°, \text{ or } 2\cos 160°, \text{ or } 2\cos 280°.$

∴ $\quad x = y - 2 = -2 + 2\cos 40°, \text{ or } -2 -2\cos 20°, \text{ or } -2 + 2\cos 80°.$

On referring to the tables we then have the values of x.

EXAMPLES XXIII

Solve the equations.

1. $2x^3 - 3x - 1 = 0$

2. $x^3 + 3x^2 - 1 = 0$

3. $x^3 - 24x - 32 = 0$

4. $x^3 - 6x^2 + 6x + 8 = 0$

5. $x^3 - 21x + 7 = 0$

6. $x^3 + 4x^2 + 2x - 1 = 0$

7. $x^3 - 7x + 5 = 0$

Maximum and Minimum Values

▨ **152.** In Art. 133, Part I, we have given one example of the maximum value of a trigonometrical expression.

We add another example.

If x and y be two positive angles whose sum is a constant angle α ($\not> \pi$), *find when* $\sin x \sin y$ *is a maximum, and extend the theorem to more than two angles.*

We have
$$2 \sin x \sin y = 2 \sin x \sin (\alpha - x)$$
$$= \cos (\alpha - 2x) - \cos \alpha.$$

Hence $2 \sin x \sin y$ is greatest when $\cos (\alpha - 2x)$ is greatest, *i.e.*, when $\alpha = 2x$, and therefore

$$x = y = \frac{\alpha}{2}.$$

The product is therefore greatest when the angles x and y are equal.

Let there be three angles x, y, and z whose sum is equal to a constant angle $\beta (\not> \pi)$. If, in the product

$$\sin x \sin y \sin z,$$

any two of the angles x and y be unequal, we can, by the preceding part of the article, increase the product by substituting for both x and y half their sum without increasing or diminishing the sum of the angles.

Hence so long as the angles x, y, and z are unequal, we can increase the given product by thus making the angles approach to equality.

The maximum value will therefore be obtained when the angles x, y, and z are equal.

This argument can clearly be applied whatever be the number of the angles, x, y, z ....

▨ **153.** We can now show that *the maximum triangle that can be inscribed in a given circle is equilateral.*

For, if R be the radius of the circle, we have (as in Ex. xxxvi. 10, Part I.) the area of the triangle

$$= 2R^2 \sin A \sin B \sin C,$$

where $A + B + C = 2\pi$, a constant angle. By the preceding article it follows that the triangle is greatest when

$$A = B = C.$$

▨ **154.** EXAMPLE. *Find the minimum positive value of the quantity* $a^2 \tan x + b^2 \cot x$.

Let $a^2 \tan x + b^2 \cot x = y,$

so that $a^2 \tan^2 x - y \cdot \tan x + b^2 = 0.$

Solving this quadratic equation, we have

$$\tan x = \frac{y \pm \sqrt{y^2 - 4a^2 b^2}}{2a^2}.$$

Since tan x is real, the quantity under the radical sign must be positive, so that y^2 must be $> 4a^2 b^2$.

Hence the least value of y is $2ab$, and the corresponding value of tan x is $\dfrac{b}{a}$.

EXAMPLES XXIV

1. If $x + y$ be a given angle, less than π, prove that

(1) sin x + sin y, and (2) cos x cos y

both have their greatest values when $x = y$.

2. If $x + y$ be a given angle, $< \dfrac{\pi}{2}$, prove that both cos x + cos y and $\cos^2 x + \cos^2 y$ have

their greatest values when $x = y$.

Find the greatest, or least, values of

3. $\dfrac{2\cos\theta}{\sqrt{3}} + \dfrac{\sqrt{3}}{2\cos\theta}$

4. $a \sec \theta - b \tan \theta$

5. $\dfrac{\cosec^2 \theta - \cot\theta}{\cosec^2 \theta + \cot\theta}$

6. $a^2 \sin^2 \theta + b^2 \cosec^2\theta$

7. $a^2 \sec^2 \theta + b^2 \cosec^2\theta$

If $x + y$ be equal to a given angle 2α, which is less than π, find the minimum value of

8. tan x + tan y

9. sec x + sec y

We can easily prove that

$$\sec x + \sec y = \cos\alpha \left[\frac{1}{\cos(\alpha - x) - \sin\alpha} + \frac{1}{\cos(\alpha - x) + \sin\alpha} \right].$$

10. If $x + y = \alpha$, where α is $\not> \dfrac{\pi}{2}$, find when tan x tan y is a maximum.

$$\left[\text{We have } 1 - \tan x \tan y = \frac{2\cos\alpha}{\cos\alpha + \cos(\alpha - 2x)} \right].$$

11. Prove that the maximum triangle having a given perimeter is equilateral.

$$\left[\text{The area of a triangle can be proved to equal } s^2 \tan\frac{A}{2} \tan\frac{B}{2} \tan\frac{C}{2} \right].$$

12. If x, y, z ... be angles whose sum is equal to a given angle, and if each of the angles be positive and less than a right angle, prove that the product cos x cos y cos z ...... is greatest when the angles are equal.

13. If ABC be a triangle, prove that the quantities $\sin A + \sin B + \sin C$ and $\sin A \sin B \sin C$ have their greatest values when the triangle is equilateral.

14. Prove that the area of the pedal triangle of an acute-angled triangle is never greater than one quarter of the area of the latter.

15. If ABC be a triangle, prove that the least value of

$$\cos 2A + \cos 2B + \cos 2C \text{ is } -\frac{3}{2}.$$

Prove also that $\cos A + \cos B + \cos C$ is always > 1 and not greater than $\frac{3}{2}$.

16. If ABC be a triangle, prove that the quantities

$\cot A + \cot B + \cot C$ and $\cot^2 A + \cot^2 B + \cot^2 C$

both have their least value when the triangle is equilateral.

On the geometrical representation of complex quantities

▨ **155.** In Chap. 4, Part I, we pointed out that if a distance in any direction (say, horizontally towards the right) be represented by a, then $-a$ represents the same distance drawn in an opposite direction, *i.e.*, horizontally towards the left.

The effect of prefixing $-$ to a is therefore (Fig. Art. 48, Part I.) to rotate OA in the positive direction through two right angles. The operation -1 performed on a therefore means turning a through two right angles.

▨ **156.** Now $\sqrt{-1} \times \sqrt{-1} = -1$; hence whatever meaning we give to the operation $\sqrt{-1}$ it must be such that *performing that operation twice shall be the same thing as performing the operation* -1.

Let us therefore assign to the operation $\sqrt{-1}$ the turning any length through one right angle in the positive direction. Performing the operation $\sqrt{-1}$ on a twice will therefore, as it should do, turn a through *two* right angles.

Hence, with this interpretation, $\sqrt{-1}\ a$ means a line drawn at right angles to the line denoted by a.

▨ **157.** We can now show what is denoted by

$$x + \sqrt{-1}\ y.$$

Draw OX and OY two lines at right angles. Measure along OX a distance OM equal to x and then draw MP parallel to OY and equal to y, so that MP represents $\sqrt{-1}\ y$. Then P is the point that represents the quantity $x + \sqrt{-1}\ y$, or, again, we may say that OP is the line representing this quantity.

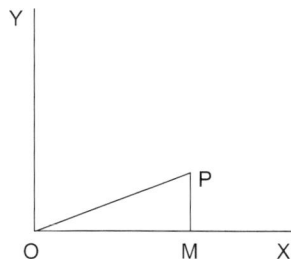

We have
$$OP = \sqrt{OM^2 + MP^2} = \sqrt{x^2 + y^2},$$

and
$$\angle MOP = \tan^{-1}\frac{MP}{OM} = \tan^{-1}\frac{y}{x}.$$

Hence the length of OP represents the modulus and MOP the principal value of the Amplitude of $x + iy$. (Art. 18.)

■ **158. Addition of two complex quantities.**

Let OP represent the complex quantity $x + iy$ and OQ represent $u + iv$, so that

$$ON = x, \ NP = y, \ OM = u,$$

and $\qquad\qquad MQ = v.$

Complete the parallelogram $OPRQ$, and draw RL perpendicular to OX and PS perpendicular to RL.

Since PR is equal and parallel to OQ, we have

$$NL = PS = OM, \text{ and } SR = MQ.$$

Hence $\qquad\quad OL = ON + NL = x + u,$

and $\qquad\quad LR = LS + SR = y + v.$

Therefore OR represents the complex quantity

$$x + u + i(y + v),$$

so that the sum of two complex quantities is represented by the diagonal of the parallelogram whose two adjacent sides represent the two given complex quantities.

■ **159.** Let

$$x + iy = r(\cos \theta + i \sin \theta),$$

as in Art. 18.

Then

$$(\cos \alpha + i \sin \alpha)(x + iy) = r(\cos \alpha + i \sin \alpha)(\cos \theta + i \sin \theta)$$
$$= r[\cos (\alpha + \theta) + i \sin (\alpha + \theta)] \qquad\qquad \text{...(1)}$$

Now, $\qquad\qquad\qquad r [\cos \theta + i \sin \theta]$

means, with our interpretation, a line of length r drawn at an angle θ with OX.

Also $\qquad\qquad r [\cos (\alpha + \theta) + i \sin (\alpha + \theta)]$

means a line of the same length r drawn at an angle $\alpha + \theta$ with OX. (Art. 157).

Hence, by (1), the effect of multiplying $x + iy$ by $\cos \alpha + i \sin \alpha$ is to turn through an angle α the line that represents $x + iy$.

■ **160.** *Geometrical meaning of De Moivre's Theorem.*

The quantity $(\cos \alpha + i \sin \alpha)(\cos \beta + i \sin \beta)(\cos \gamma + i \sin \gamma)(\cos \delta + i \sin \delta)$ means the line represented by $\cos \delta + i \sin \delta$ turned first through an angle γ, then through β, and finally through a_x, *i.e.*, altogether turned through $\alpha + \beta + \gamma$.

But this total operation gives the same line as

$$[\cos (\alpha + \beta + \gamma) + i \sin (\alpha + \beta + \gamma)][\cos \delta + i \sin \delta].$$

Similarly for any number of factors.

Hence De Moivre's Theorem expresses algebraically the geometrical fact that to turn a line through a number of angles successively has the same effect as turning the line through an angle equal to the sum of the angles.

■ **EXAMPLE.** *The three cube roots of unity are easily found to be*

$$\cos 0 + i \sin 0, \ \cos \frac{2\pi}{3} + i \sin \frac{2\pi}{3},$$

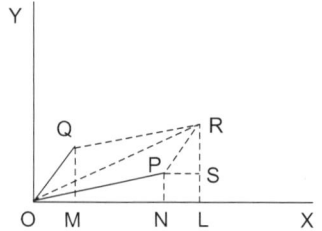

and
$$\cos\frac{4\pi}{3} + i\sin\frac{4\pi}{3},$$

so that we have

$$(\cos 0 + i \sin 0)(\cos 0 + i \sin 0)(\cos 0 + i \sin 0) = 1,$$

$$\left(\cos\frac{2\pi}{3} + i\sin\frac{2\pi}{3}\right)\left(\cos\frac{2\pi}{3} + i\sin\frac{2\pi}{3}\right)\left(\cos\frac{2\pi}{3} + i\sin\frac{2\pi}{3}\right) = 1,$$

and
$$\left(\cos\frac{4\pi}{3} + i\sin\frac{4\pi}{3}\right)\left(\cos\frac{4\pi}{3} + i\sin\frac{4\pi}{3}\right)\left(\cos\frac{4\pi}{3} + i\sin\frac{4\pi}{3}\right) = 1.$$

The first of these equations states that turning a line three times in succession through a zero angle gives the original line.

The second states that turning it three times in succession through an angle $\frac{2\pi}{3}$, (*i.e.* altogether through 2π) gives the original line.

The third states that turning it three times in succession through an angle $\frac{4\pi}{3}$, (*i.e.* altogether through 4π) gives the original line.

These statements are all clearly true.

▪ **161. Multiplication of two complex quantities.**

If
$$x + iy = r (\cos \theta + i \sin \theta),$$
and
$$u + iv = \rho(\cos \phi + i \sin \phi),$$
we have

$$(u + iv)(x + iy) = r_p[\cos (\theta + \phi) + i \sin (\theta + \phi)].$$

The effect of multiplying a complex quantity $x + iy$ by another $u + iv$ is therefore to turn the line representing $x + iy$ through an angle

$$\phi\left[\text{i.e., } \tan^{-1}\frac{v}{u}\right],$$

and to alter its length in the ratio

$$1 : p, \text{ i.e. } 1 : \sqrt{u^2 + v^2}.$$

Hence the multiplying of one complex quantity by another is represented by "a turning and a stretching."

MISCELLANEOUS EXAMPLES XXV

1. Prove that the equation $\tan x = kx$ has an infinite number of real roots.
2. If A, B and C be the angles of a triangle, prove that
$$1 - 8 \cos A \cos B \cos C$$
   is always positive.
3. If α and β be the imaginary cube roots of unity prove that

$$\alpha e^{\alpha x} + \beta e^{\beta x} = -e^{-\frac{x}{2}}\left[\frac{\sqrt{3}}{2}\sin\frac{\sqrt{3}x}{2} + \cos\frac{\sqrt{3}x}{2}\right].$$

4. If x be less than a radian prove that $x = 2\sqrt{\dfrac{3-3\cos x}{5+\cos x}}$ very nearly, the error in the left-hand member being nearly $\dfrac{x^3}{480}$ radians.

5. If $\cos(\theta + i\phi) = \sec(\alpha + i\beta)$, where α, β, θ, and ϕ are all real, prove that $\tanh^2 \phi \cosh^2 \beta = \sin^2 \alpha$ and $\tanh^2 \beta \cosh^2 \phi = \sin^2 \theta$.

6. If $\qquad x = 2\cos\alpha\cosh\beta$ and $y = 2\sin\alpha\sinh\beta$,

   prove that

$$\sec(\alpha + i\beta) + \sec(\alpha - i\beta) = \frac{4x}{x^2 + y^2},$$

   and $\qquad \sec(\alpha + i\beta) - \sec(\alpha - i\beta) = \dfrac{4iy}{x^2 + y^2}.$

7. Prove that
$$\sin^n \phi \cos n\theta + n\sin^{n-1} \phi \cos(n-1)\theta \sin(\theta - \phi)$$

$$+\frac{n(n-1)}{1\cdot 2}\sin^{n-2} \phi\cos(n-2)\theta\sin^2(\theta - \phi) + \dots + \sin^n(\theta - \phi) = \sin^n\theta \cos n\phi.$$

8. Prove that the roots of the equation

$$x^n \sin n\theta - nx^{n-1} \sin(n\theta + \phi) + \frac{n(n-1)}{1\cdot 2}x^{n-2}\sin(n\theta + 2\phi) - \dots \text{ to } (n+1) \text{ terms} = 0,$$

   are given by $\qquad x = \sin\left(\theta + \phi - k\dfrac{\pi}{n}\right)\mathrm{cosec}\left(\theta - k\dfrac{\pi}{n}\right),$

   where n is an integer and k has any integral value from 0 to $n - 1$.

9. Prove that the sum to infinity of the series

$$\sin\theta + \frac{1}{2}\frac{\sin^3\theta}{3} + \frac{1\cdot 3}{2\cdot 4}\frac{\sin^5\theta}{5} + \dots$$

   is θ, if θ be acute, and, generally, is $n\pi + (-1)^n\,\theta$, where n is so chosen that $n\pi + (-1)^n$

   θ lies between $-\dfrac{\pi}{2}$ and $+\dfrac{\pi}{2}$.

10. If the arc of a circle of radius unity be divided into n equal arcs, and right-angled isosceles triangles be described on the chords of these arcs as hypotenuses and have their vertices outwards, prove that when n is indefinitely increased the limit of the product of the distances of the vertices from the centre is $e^{n/2}$, where α is the angle subtended by the arc at the centre.

11. The sides of a regular polygon of n sides, which is inscribed in a circle, meet the tangent at any point P of the circle in A, B, C, D ... Prove that the product $PA \cdot PB \cdot PC \cdot PD \dots = a^n \tan n^0$ or a^n, according as n is odd or even, where a is the radius of the circle and θ is the angle which the line joining P to an angular point subtends at the circumference.

12. A regular polygon of n sides is inscribed in a circle and from any point in the circumference chords are drawn to the angular points; if these chords be denoted by $c_1, c_2, \dots, c_n$, beginning with the chord drawn to the nearest angular point and taking the rest in order, prove that the quantity

$$c_1c_2 + c_2c_3 + \dots + c_{n-1}c_n - c_nc_1$$

is independent of the position of the point from which the chords are drawn.

13. A series of radii divide the circumference of a circle into $2n$ equal parts; prove that the product of the perpendiculars let fall from any point of the circumference upon n successive radii is

$$\frac{r^n}{2^{n-1}}\sin n\theta,$$

where r is the radius of the circle and θ is the angle between one of the extreme of these radii and the radius to the given point.

14. If a regular polygon of n sides be inscribed in a circle, and l be the length of the chord joining any fixed point on the circle to one of the angular points of the polygon, prove that

$$\Sigma l^{2m} = na^{2m}\frac{\lfloor 2m}{\{\lfloor m\}^2}.$$

15. $ABCD...$ is a regular polygon of n sides which is inscribed in a circle, whose radius is a and whose centre is O; prove that the product of the distances of its angular points from a straight line at right angles to OA and at a distance $b\ (> a)$ from the centre is

$$b^n\left[\cos^n\left(\frac{1}{2}\sin^{-1}\frac{a}{b}\right) - \sin^n\left(\frac{1}{2}\sin^{-1}\frac{a}{b}\right)\right]^2.$$

16. Prove that there is one, and only one, solution of the equation $\theta = \cos\theta$ and that it is less than $\dfrac{\pi}{4}$.

17. Prove that the general value of θ which satisfies the equation

$$(\cos\theta + i\sin\theta)(\cos 2\theta + i\sin 2\theta)... \text{ to } n \text{ factors} = 1$$

is $\dfrac{4m\pi}{n(n+1)}$, where m is any integer.

18. Prove that

$$e^{\pi} + e^{-\pi} = 2\{1 + 2^2\}\left\{1 + \left(\frac{2}{3}\right)^2\right\}\left\{1 + \left(\frac{2}{5}\right)^2\right\}\dots \text{ ad inf.}$$

19. Prove that

$$1 + \frac{x^3}{\lfloor 3} + \frac{x^6}{\lfloor 6} + \frac{x^9}{\lfloor 9} + \dots\text{ad inf.} = \frac{1}{3}\left[e^x + 2e^{-\frac{x}{2}}\cos\left(\frac{\sqrt{3}x}{2}\right)\right].$$

20. Show that

$$x + \frac{x^4}{\lfloor x} + \frac{x^7}{\lfloor 7} + \frac{x^{10}}{\lfloor 10}\ \lfloor \dots\dots \text{ ad inf.}$$

$$= \frac{1}{3}e^x - \frac{1}{3}e^{-x/2}\left(\cos\frac{x\sqrt{3}}{2} - \sqrt{3}\sin\frac{x\sqrt{3}}{2}\right).$$

21. Show that the sum of the series

$$\sum_{r=1}^{r=\infty}\left[\frac{1}{(3r-1)^4} + \frac{1}{(3r+1)^4}\right] \text{ is } \frac{8\pi^4}{729} - 1.$$

22. Prove that

$$\cos\frac{2\pi}{17} + \cos\frac{4\pi}{17} + \cos\frac{6\pi}{17} + ... + \cos\frac{14\pi}{17} + \cos\frac{16\pi}{17} = -\frac{1}{2},$$

and $\sec\dfrac{2\pi}{17} + \sec\dfrac{4\pi}{17} + ... + \sec\dfrac{14\pi}{17} + \sec\dfrac{16\pi}{17} = 8.$

23. If $\alpha = \dfrac{2\pi}{21}$, prove that the values of

$$\cos\alpha + \cos 5\alpha + \cos 17\alpha$$

and $$\cos 11\alpha + \cos 13\alpha + \cos 19\alpha$$

are respectively $\dfrac{\sqrt{21}+1}{4}$ and $-\dfrac{\sqrt{21}+1}{4}$.

24. Prove that

$$\tan\alpha + \tan\left(\alpha + \frac{2\pi}{5}\right) + \tan\left(\alpha + \frac{4\pi}{5}\right) + \tan\left(\alpha + \frac{6\pi}{5}\right) + \tan\left(\alpha + \frac{8\pi}{5}\right) = 5\tan 5a.$$

25. Show that the equation whose roots are $\tan\dfrac{r\pi}{15}$, where r is any number including unity less than and prime to 15, is
$$x^8 - 92x^6 + 134x^4 - 28x^2 + 1 = 0.$$

26. From the sum of the series

$$\sin 2\theta - \frac{1}{2}\sin 4\theta + \frac{1}{3}\sin 6\theta - ... \text{ ad inf.,}$$

or otherwise, show that

$$\frac{\pi\sqrt{2}}{4} = 1 + \frac{1}{3} - \frac{1}{5} - \frac{1}{7} + \frac{1}{9} + \frac{1}{11} - ... \text{ ad inf.}$$

27. Assuming equation (4) of Art. 53, show that

$$\theta^2 = \sin^2\theta + \frac{2}{3}\frac{\sin^4\theta}{2} + \frac{2\cdot 4}{3\cdot 5}\frac{\sin^6\theta}{3} + ...$$

28. Prove that

$$\frac{1}{2x}\frac{\sinh x}{\cosh x - \cos\alpha} - \frac{1}{a^2 + x^2} = \sum_{n=1}^{n=\infty}\left\{\frac{1}{(2n\pi - a)^2 + x^2} + \frac{1}{(2n\pi + \alpha)^2 + x^2}\right\}.$$

29. Prove that the general value of $\sinh^{-1}x$ is

$$ik\pi + (-1)^k \log\left[x + \sqrt{1+x^2}\right],$$

where k is any integer.

30. The side BC of a square $ABCD$ is produced indefinitely, and along it are measured CC_1, C_1C_2, C_2C_3, ... each equal to BC.

If $\theta_1, \theta_2, \theta_3, ...$ be the angles BAC_1, BAC_2, BAC_3, ..., prove that

$$\sin\theta_1 \sin\theta_2 \sin\theta_3 ... \text{ ad. inf.} = 2\sqrt{\frac{\pi}{e^\pi - e^{-\pi}}}.$$

31. If $\rho_1, \rho_2, \ldots \rho_n$ be the distances of the vertices of a regular polygon of n sides from any point P in its plane, prove that

$$\frac{1}{\rho_1^2} + \frac{1}{\rho_2^2} + \ldots\ldots + \frac{1}{\rho n^2} = \frac{n}{r^2 - a^2} \frac{r^{2n} - a^{2n}}{r^{2n} - 2a^n r^n \cos n\theta + a^{2n}},$$

where a is the radius of the circumcircle of the polygon, r is the distance of P from its centre O, and θ is the angle that OP makes with the radius to any angular point of the polygon.

32. If $\theta + \phi + \psi = 2\pi$, prove that

$$\cos^2\theta + \cos^2\phi + \cos^2\psi - 2\cos\theta\cos\phi\cos\psi = 1.$$

Hence deduce the relation between the lengths of the six straight lines joining four points which are in one plane.

33. Show that the general value of $\log(-1)$ is $(2n+1)\pi i$, and point out the fallacy in the following:

$$\log_e(-1) = 1/2\log_e(-1)^2 = 1/2\log_e 1 = 0;$$
$$\therefore\ -1 = e^0 = 1.$$

34. Prove that the series $\sum\limits_{n=0}^{n=\infty} x^n \sinh(n+1)\alpha$ is convergent if x is numerically less than $e^{-\alpha}$, α being assumed to be positive, and that the sum is $\sinh\alpha/(1 - 2x\cosh\alpha + x^2)$; but that the series $\sum\limits_{n=0}^{n=\infty} x^n \sin(n+1)\alpha$ is convergent provided that x is numerically less than unity, the sum being $\sin\alpha/(1 - 2x\cos\alpha + x^2)$.

35. Assuming the formula for $\sin\theta$ in factors, prove that

$$(1+x)\left(1 - \frac{x}{5}\right)\left(1 + \frac{x}{7}\right)\left(1 - \frac{x}{11}\right)\left(1 + \frac{x}{13}\right)\ldots = \cos\frac{\pi x}{6} + \sqrt{3}\sin\frac{\pi x}{6},$$

where the signs alternate in the factors and the denominators are the odd integers not divisible by 3 in ascending order.

Show that $\qquad 1 - \dfrac{1}{5^3} + \dfrac{1}{7^3} - \dfrac{1}{11^3} + \ldots$ to $\infty = \dfrac{\pi^3\sqrt{3}}{54}.$

36. A point is taken in the plane of a regular polygon of n sides at a distance c from the centre and on the line joining the centre to a vertex, and the radius of the inscribed circle is r. Show that the product of the distances of the point from the sides of the polygon is

$$\frac{c^n}{2^{n-2}}\cos^2\left(\frac{\pi}{2}\cos^{-1}\frac{r}{c}\right),\ \text{if}\ c > r,$$

and is $\qquad \dfrac{c^n}{2^{n-2}}\cosh^2\left(\dfrac{n}{2}\cosh^{-1}\dfrac{r}{c}\right),\ \text{if}\ c < r.$

ADDITIONAL MISCELLANEOUS EXAMPLES

1. If $(a_1 + b_1 i)(a_2 + b_2 i)\ldots(a_n + b_n i) = A + Bi$, prove that

$$\tan^{-1}\frac{b_1}{a_1} + \tan^{-1}\frac{b_2}{a_2} + \ldots + \tan^{-1}\frac{b_n}{a_n} = \tan^{-1}\frac{B}{A}.$$

[*Use De Moivre's Theorem.*]

2. When x is small, show that

$$\log \sin x = \log x - \frac{x^2}{6} - \frac{x^4}{180}.$$

3. Show that

$$\sinh(\beta - \gamma) + \sinh(\gamma - \alpha) + \sinh(\alpha - \beta) = 4 \sinh$$

$$= 4 \sinh \frac{\beta - \gamma}{2} \sinh \frac{\gamma - \alpha}{2} \sinh \frac{\alpha - \beta}{2}.$$

4. Prove that the circular measure of an angle θ is equal to the sum of a constant and one of the two series

$$\tan \theta - \frac{1}{3} \tan^3 \theta + \frac{1}{5} \tan^5 \theta - ......,$$

$$- \cot \theta + \frac{1}{3} \cot^3 \theta - \frac{1}{5} \cot^5 \theta + ...,$$

distinguishing the cases.

Give the constants for the angles $49°$ and $200°$.

5. Sum the series

$$1 - \frac{x^2}{\underline{|2}} + \frac{3x^4}{\underline{|4}} - \frac{5x^6}{\underline{|6}} + ... \text{ to infinity.}$$

6. Prove that

$$\coth^{-1} x = \frac{1}{2} \log \frac{x+1}{x-1}.$$

Expand $\coth^{-1} x$ in a series of powers of x.

7. Show that

$$\frac{1}{2\cos\alpha} - \frac{1}{2\cos\alpha} - \frac{1}{2\cos\alpha} - ... - \frac{1}{2\cos\alpha + p}$$

$$= \frac{\sin n\alpha + p\sin(n-1)\alpha}{\sin(n+1)\alpha + p\sin n\alpha},$$

there being n quotients on the left-hand side.

[*Use the method of Induction.*]

8. Show that the geometric mean of the cosines of n acute angles is never greater than the cosine of the arithmetic mean of the angles.

9. Find all the cube roots of $88 + 16\sqrt{-1}$ having given that, when $\tan \theta = 2$, then

$$\tan 3\theta = \frac{2}{11}.$$

10. Find the limit to which

$$\frac{e^{\sin x} - e^{-\sin x} - 2\tan x}{\tan x - x}$$

tends as x tends towards zero.

11. Prove that

(i) $2\tan^{-1}\left[\sqrt{\dfrac{a-b}{a+b}}\,\tan\dfrac{x}{2}\right] = \cos^{-1}\dfrac{b+a\cos x}{a+b\cos x};$

(ii) $\log\dfrac{\sqrt{b+a}+\sqrt{b-a}\,\tan\dfrac{x}{2}}{\sqrt{b+a}-\sqrt{b-a}\,\tan\dfrac{x}{2}} = \cosh^{-1}\dfrac{b+a\cos x}{a+b\cos x}.$

12. Find the sum of the series

$$\frac{2}{1\cdot 3}\sin 2x - \frac{4}{3\cdot 5}\sin 4x + \frac{6}{5\cdot 7}\sin 6x - \text{...ad inf.}$$

or all values of x between 0 and π.

13. Prove that the sum of $n-1$ terms of the series

$\qquad \tan\alpha \tan 2\alpha + \tan 2\alpha \tan 3\alpha + \tan 3\alpha \tan 4\alpha$ ...

is equal to $\tan n\alpha \cot\alpha - n$.

Deduce the sum of the series

$1.2 + 2.3 + 3.4 + \ldots$ to $(n-1)$ terms.

14. Prove that

$$\sqrt{2} = \frac{4\cdot 36\cdot 100\cdot 196\cdot 324...}{3\cdot 35\cdot 99\cdot 195\cdot 323...},$$

and that $\qquad \dfrac{\sqrt{3}}{2} = \dfrac{8\cdot 80\cdot 224\cdot 440...}{9\cdot 81\cdot 225\cdot 441...}.$

[*Use the result of Art. 123.*]

15. A triangular piece of ground is surveyed and the sides measured as 200, 300 and 400 m. respectively; but the measuring chain has worn so that its real length is 2% greater than its nominal length. Find the error in the calculated area of the triangle.

16. In any triangle, show that the radius of the inscribed circle is never greater than half the radius of the circumscribed circle.

[*Use the Corollary of Art. 204, Part* I.]

---

17. If $x = \cos\theta + \sqrt{-1}\sin\theta$, prove that

$$\frac{1}{2x} + \frac{1}{2\pi} + \frac{1}{2x} + \ldots\ldots \text{ ad inf.}$$

$$= (\cos\theta + \cos^2\theta)^{1/2} - \cos\theta + \sqrt{-1}\,[(\cos\theta - \cos^2\theta)^{1/2} - \sin\theta].$$

18. Prove Snellius' formula, that x differs from

$$\frac{3\sin 2x}{2(2+\cos 2x)} \text{ by } \frac{4x^5}{45} \text{ nearly,}$$

when x is small.

19. Show that

$$\tan^{-1}\left(i\,\frac{x-a}{x+a}\right) = -\frac{i}{2}\log\frac{a}{x}.$$

20. Sum to infinity the series

$$\frac{7}{1 \cdot 3 \cdot 5} + \frac{19}{5 \cdot 7 \cdot 9} + \frac{31}{9 \cdot 11 \cdot 13} + \dots;$$

the numerators being in arithmetical progression.

[*Put* $\theta = \dfrac{\pi}{4}$ *in the result of Art.* 94.]

21. Sum to infinity the series

$$\frac{\cos\theta}{1 \cdot 2} + \frac{\cos 2\theta}{2 \cdot 3} + \frac{\cos 3\theta}{3 \cdot 4} + \dots$$

22. Sum to n terms the series

$$\tan\alpha + 2\tan 2\alpha + 2^2 \tan 2^2 \alpha.$$

23. Expand $\dfrac{\sin\theta}{1 - \sin\alpha\cos\theta}$ in a series of sines of multiples of θ.

24. Sum of the series

$$\sum_{n=-\infty}^{n=\infty} \frac{An + B}{(n+a)(n+b)(n+c)}.$$

[*Break into partial fractions and use Page* 158, *Ex.* 7.]

25. If $\cos\theta + \cos\phi + \cos\psi = 0$, and $\sin\theta + \sin\phi + \sin\psi = 0$,

then $\qquad \cos 3\theta + \cos 3\phi + \cos 3\psi - 3\cos(\theta + \phi + \psi) = 0$,

and $\qquad \sin 3\theta + \sin 3\phi + 3\sin\psi - 3\sin(\theta + \phi + \psi) = 0$.

26. Show that the angle whose sine is $\dfrac{1}{4}\sqrt{3}$ differs from the seventh part of two right angles by less than the thousandth part of a radian.

27. Show that the area of a segment of a circle of height h, bounded by a chord of length c, is $\dfrac{2}{3}hc$, if powers of $\dfrac{h}{c}$ above the first be neglected.

28. Show that the solutions of the equation $\sinh x - \sinh\alpha$ are all included in the expression

$$n\pi i + (-i)^n \alpha.$$

29. If x lies between 0 and 2π, prove that

$$\frac{\sin 2x}{1 \cdot 3} + \frac{\sin 3x}{2 \cdot 4} + \frac{\sin 4x}{3 \cdot 5} + \dots \text{ ad inf.}$$

$$= \frac{1}{4}\sin x \left[1 - 4\log\left(2\sin\frac{x}{2}\right)\right].$$

30. Given that $\tan(\phi + \theta)\cos 2\alpha = \tan\phi$,
prove that

$$\theta = \tan^2\alpha\sin 2\phi + \frac{1}{2}\tan^4\alpha\sin 4\theta + \frac{1}{3}\tan^6\alpha\sin 6\phi + \dots$$

31. The area of a triangle is determined by measurements which give $b = 125$ m, $c = 160$ m, $A = 57°35'$. Another set of measurements give $b_1 = 125.5$ m, $c_1 = 161$ m, $A_1 = 57° 25'$. Find the percentage difference between the second determination of the area and the first.

32. Find the maximum value of

$$\sin (\alpha - \beta) + \sin (\beta - \gamma) + \sin (\gamma - \alpha).$$

[*Consider the maximum area of triangle ABC inscribed in a circle of centre O, where OA, OB, OC make angles* α, β, γ *with a fixed line.*]

---

33. From the identity

$$\frac{a^2}{(a-b)(a-c)} + \frac{b^2}{(b-a)(b-c)} + \frac{c^2}{(c-a)(c-b)} = 1,$$

deduce the identities

$$\Sigma \cos 3(\alpha + \theta) \sin (\beta - \gamma)$$

$$= 4 \cos (3\theta + \alpha + \beta + \gamma) \sin (\beta - \gamma) \sin (\gamma - \alpha) \sin (\alpha - \beta),$$

and $\Sigma \sin 3(\alpha + \theta) \sin (\beta - \gamma)$

$$= 4 \sin (3\theta + \alpha + \beta + \gamma) \sin (\beta - \gamma) \sin (\gamma - \alpha) \sin (\alpha - \beta).$$

[*Put* $a = \cos (2\alpha + 2\theta) + i \sin (2\alpha + 2\theta)$ *etc.*]

34. Prove that $\tan x - 24 \tan \dfrac{x}{2}$ differs from $4 \sin x - 15 x$ by a quantity of the seventh order at least.

35. If a be the length of the chord of a circular arc, and b that of the chord of half the arc, prove that the length of the arc is

$$\frac{8b - a}{3} \text{ approximately.}$$

If c be the length of the chord of one quarter of the arc, prove that a nearer approximation is

$$\frac{a - 40b + 256c.}{45}$$

If the arc be a quadrant, show that these approximations give the value of π correct to 2 and 5 places of decimals respectively.

36. If $\log_e \log_e (x + iy) = p + iq$, then

$$y = x \tan \left\{ \tan q \log_e \sqrt{x^2 + y^2} \right\}.$$

37. Prove that

$$\frac{\cos\theta - \dfrac{\cos^3 \theta}{\underline{3}} + \dfrac{\cos^5 \theta}{\underline{5}} - \dots}{1 - \dfrac{\cos^2 \theta}{\underline{2}} + \dfrac{\cos^4 \theta}{\underline{4}} - \dots} = \frac{\cos\theta - \dfrac{\cos 3\theta}{\underline{3}} + \dfrac{\cos 5\theta}{\underline{5}} - \dots}{1 - \dfrac{\cos 2\theta}{\underline{2}} + \dfrac{\cos 4\theta}{\underline{4}} - \dots}$$

38. Find the sum of the series

$$\sin \theta \sec 3\theta + \sin 3\theta \sec 3^2\theta + \sin 3^2\theta \sec 3^3\theta + \dots \text{ to } n \text{ terms.}$$

39. In a triangle ABC, if $b < a$, show that

(*i*) $B = \dfrac{b}{a}\sin C + \dfrac{1}{2}\dfrac{b^2}{a^2}\sin 2C + \dfrac{1}{3}\dfrac{b^3}{a^3}\sin 3C + \dots,$

and (*ii*) $\dfrac{a^n}{c^n}\sin nB = n\dfrac{b}{a}\sin C + \dfrac{n(n+1)}{1 \cdot 2}\dfrac{b^2}{a^2}\sin 2C + \dfrac{n(n+1)(n+2)}{1 \cdot 2 \cdot 3}\dfrac{b^3}{a^3}\sin 3C + \dots$

40. The sides of a triangle are observed to be $a = 2$, $b = 3$, $c = 4$, but it is known that there is a small error in the measurement of c; find which angle can be found with the greatest accuracy.

---

41. From the identity

$$a^2 \frac{(x-b)(x-c)}{(a-b)(a-c)} + b^2 \frac{(x-c)(x-a)}{(b-c)(b-a)} + c^2 \frac{(x-a)(x-b)}{(c-a)(c-b)} = x^2,$$

obtain the identities

$$\cos 2(\theta + \alpha)\frac{\sin(\theta - \beta)\sin(\theta - \gamma)}{\sin(\alpha - \beta)\sin(\alpha - \gamma)} + \text{two similar terms} = \cos 4\theta,$$

and $\sin 2(\theta - \alpha)\dfrac{\sin(\theta - \beta)\sin(\theta - \gamma)}{\sin(\alpha - \beta)\sin(\alpha - \gamma)} + \text{two similar terms} = \sin 4\theta.$

42. If $\tan(\theta - \phi) = \dfrac{e^2 \sin\theta\cos\theta}{1 - e^2 \sin^2\theta}$, and e be small, prove that

$$\phi = \theta - \frac{e^2}{2}\sin 2\theta - \frac{e^4}{8}(2\sin 2\theta - \sin 4\theta) + \ldots$$

43. If $\qquad \cos\left(\dfrac{\pi}{2}\sin\theta\right) = \sin\left(\dfrac{\pi}{2}\cos\theta\right),$

show that θ has four sets of values given by

$$\pm\theta = \left(2m + \frac{3}{4}\right)\pi \pm i\log\frac{4n - 1 + \sqrt{16n^2 - 8n - 1}}{\sqrt{2}},$$

where m and n are any integers, positive or negative, and m may be also zero. What is the solution when $n = 0$?

44. Find the sum to n terms of the series

$$\tan\theta\tan^2\frac{\theta}{2} + 2\tan\frac{\theta}{2}\tan^2\frac{\theta}{2^3} + 2^2\tan\frac{\theta}{2^2}\tan^2\frac{\theta}{2^3} + \ldots\ldots$$

45. If $\cot\phi = \dfrac{1}{x} + \cot\theta$, expand ϕ in a series proceeding by ascending powers of x.

46. Show that the sum to infinity of the series

$$\frac{1\cdot 2}{3^4} + \frac{2\cdot 3}{5^4} + \frac{3\cdot 4}{7^4} + \ldots\ldots$$

is $\qquad \dfrac{\pi^2\left(12 - \pi^2\right)}{384}.$

[*Use Art.* 127.]

47. Evaluate the continued fraction

$$\cot\theta - \frac{2}{\cot 2\theta} - \frac{2}{\cot 2^2\theta} - \ldots - \frac{2}{\cot 2^{n-1}\theta} - \frac{2}{\tan^{2n}\theta}.$$

48. Prove that in any plane triangle the value of

$$\tan B \tan C + \tan C \tan A + \tan A \tan B$$

cannot lie between 0 and 9.

[*Show that the expression* $= 1 + \sec A \sec B \sec C$ *and then apply the method of Art.* 152.]

---

49. If $\cos z = \cos (z + x) \cos \Delta + \sin (z + x) \sin \Delta \cos h$, where x and Δ are so small that powers higher than their cubes may be neglected show that

$$x = \Delta \cosh h - \frac{1}{2}\Delta^2 \cot z \sin^2 h + \frac{1}{3}\Delta^3 \cos h \sin^2 h.$$

50. Prove that if

$$(1 + i \tan \alpha)^{1 + i \tan \beta}$$

can have real values, one of them is

$$\left(\sec \alpha\right)^{\sec^2 \beta}.$$

51. Sum to n terms the series

$$\frac{\cot^2 \alpha}{1 - \cos^2 2\alpha \sec^2 \alpha} + \frac{\cot^3 \alpha}{1 - \cos^2 3\alpha \sec^2 \alpha} + \frac{\cot^4 \alpha}{1 - \cos^2 4\alpha \sec^2 \alpha} + \dots$$

52. Prove that

$$\sqrt{1 + \operatorname{cosec} \frac{\theta}{2}} = \left(1 - e^{\theta i}\right)^{-\frac{1}{2}} + \left(1 - e^{-\theta i}\right)^{-\frac{1}{2}},$$

and deduce an expansion of

$$\frac{1}{2}\sqrt{1 + \operatorname{cosec} \frac{\theta}{2}}$$

in cosines of multiples of θ.

53. Prove that

$$\frac{(x+1)^{2n} + (x-1)^{2n}}{2} = \prod_{r=1}^{r=n}\left[x^2 + \tan^2 \frac{(2r-1)\pi}{4n}\right].$$

[*Use the first formula of Art.* 120]

54. Find the sum of the series

$$\frac{1}{1^2 \cdot 2^2} + \frac{1}{2^2 \cdot 3^2} + \frac{1}{3^2 \cdot 4^2} + \dots \text{ ad inf.}$$

55. Show that the area of the greatest triangle, whose base is b and the ratio of whose sides is r, is $\dfrac{rb^2}{2(r^2 - 1)}$.

56. Show by a graph that the equation $\sin x = \tanh x$ has an infinite number of real roots, and that the large positive roots occur in pairs one a little less and the other a little greater than $(2p + \frac{1}{2})\pi$, where p is a large positive integer.

---

57. Regular polygons of n sides are inscribed in and circumscribed to a circle. Show that, if n be large, by taking the mean of the perimeters we get a nearer approximation to π than we should get by taking the mean of the areas by about $\dfrac{\pi^3}{12n^2}$.

58. A regular polygon of n sides is inscribed in a circle of radius a; prove that the sum of the reciprocals of the distances of the angular points of the polygon from a tangent to the circle is

$$\frac{n^2}{2a}\operatorname{cosec}^2 n\theta,$$

where 2θ is the angle which a radius drawn to the point of contact of the tangent makes with the radius drawn to one of the angular points of the polygon.

59. Show that the principal value of

$$\frac{(a+bi)^{p+qi}}{(a-bi)^{p-qi}}$$

is $\cos 2(p\alpha + q \log r) + i \sin 2 (p\alpha + q \log r)$,

where $r = \sqrt{a^2 + b^2}$, and $\alpha = \tan^{-1}\dfrac{b}{a}$.

60. Sum to n terms the series

$$\operatorname{cosec}\theta + \operatorname{cosec}\frac{\theta}{2} + \operatorname{cosec}\frac{\theta}{2^2} + \dots$$

61. Sum to n terms the series

$$\frac{1}{1 - \tanh\alpha\tanh 2\alpha} + \frac{1}{1 - \tanh 2\alpha\tanh 4\alpha} + \frac{1}{1 - \tanh 3\alpha\tanh 6\alpha} + \dots$$

62. Show that

$$\frac{\sin\theta}{1 - 2y + y^2 \sec^2\theta} = \sum_{n=0}^{n=\infty} y^n \sec^n\theta\sin(n+1)\theta.$$

63. Prove that

$$\frac{\left(\dfrac{\pi^2}{4}+1\right)\left(\dfrac{\pi^2}{4}+\dfrac{1}{9}\right)\left(\dfrac{\pi^2}{4}+\dfrac{1}{25}\right)\dots}{\left(\dfrac{\pi^2}{4}+\dfrac{1}{4}\right)\left(\dfrac{\pi^2}{4}+\dfrac{1}{16}\right)\left(\dfrac{\pi^2}{4}+\dfrac{1}{36}\right)\dots} = \frac{e^2+1}{e^2-1}.$$

64. Prove that

$$\sin\theta + \cos\theta = \left(1+\frac{4\theta}{\pi}\right)\left(1-\frac{4\theta}{3\pi}\right)\left(1+\frac{4\theta}{5\pi}\right)\left(1-\frac{4\theta}{7\pi}\right)\dots,$$

and deduce that

$$1 - \frac{1}{3^3} + \frac{1}{5^3} - \frac{1}{7^3} + \dots = \frac{\pi^3}{32}.$$

[In the result of Exs. XXI, No. 13 put $\alpha = \dfrac{\pi}{4}$. Then take logarithms and equate the coefficients of power of θ.]

65. Show that

$$\cos^2\frac{\pi}{11} + \cos^2\frac{2\pi}{11} + \cos^2\frac{3\pi}{11} + \cos^2\frac{4\pi}{11} + \cos^2\frac{5\pi}{11} = \frac{9}{4},$$

66. If $x = \sin\frac{\pi}{9}$, show that x is a root of the equation

$$64x^6 - 96x^4 + 36x^2 - 3 = 0,$$

and write down the other roots of the equation.

67. Show that the roots of the equation

$$y^6 - y^5 - 6y^4 + 6y^3 + 8y^2 - 8y + 1 = 0$$

are the values of $2\cos\frac{r\pi}{21}$, where r has either of the values 2, 4, 8, 10, 16, 20.

68. If $x + iy = b(\cos\beta + i\sin\beta)(\cos\theta + i\sin\theta) + c(\cos\gamma + i\sin\gamma)(\cos\theta - i\sin\theta)$, where b, β, c, γ are real constants and x, y, θ real variables, show that as θ varies the point (x, y) describes a conic section.

69. If $\phi = \theta - 2e\sin\theta + \frac{3e^2}{4}\sin 2\theta - \frac{e^3}{3}\sin 3\theta$,

prove that $\theta = \phi + 2e\sin\phi + \frac{5e^2}{4}\sin 2\phi + \frac{e^3}{12}(13\sin 3\phi - 3\sin\phi)$,

where powers of e above the third are neglected.

70. Prove that, if powers of θ beyond the fifth be neglected, then

$$\theta = 2\sin\frac{\theta}{8} + \sqrt{\left(\sin\theta - 2\sin\frac{\theta}{8}\right)^2 + (1 - \cos\theta)^2}.$$

71. If A, B, C, D be four consecutive vertices of a regular heptagon inscribed in a circle of radius unity, prove that

$$AC + AD - AB = \sqrt{7}.$$

72. Prove that

$$\sec x = 1 + \frac{x^2}{\lfloor 2} + \frac{5x^4}{\lfloor 4} + \frac{61x^6}{\lfloor 6} + \frac{1385x^8}{\lfloor 8} + \dots,$$

and $\sec^2 x = 1 + \frac{3x^2}{\lfloor 2} + \frac{33x^4}{\lfloor 4} + \frac{723x^6}{\lfloor 6} + \dots$

73. If n is odd, prove that

$$\cot^2\frac{2\pi}{2n} + \cot^2\frac{4\pi}{2n} + \dots + \cot^2\frac{(n-1)\pi}{2n} = \frac{1}{6}(n-1)(n-2),$$

74. If n be even, prove that

$$\cot^2\frac{\pi}{2n} + \cot^2\frac{3\pi}{2n} + \dots + \cot^2\frac{(n-1)\pi}{2n} = \frac{1}{2}n(n-1).$$

75. If n is an even integer, prove that

$$\sec^2\frac{\pi}{2n} + \sec^2\frac{5\pi}{2n} + \sec^2\frac{9\pi}{2n} + \dots + \sec^2\frac{(2n-3)\pi}{2n} = \frac{n^2}{2}.$$

76. Prove that

$$\sum_{r=0}^{r=n-1} \sec^2\left(\alpha + \frac{2r\pi}{n}\right)$$

is equal to $n^2 \sec^2 n\alpha$, or to

$$\frac{n^2}{1-(-1)^{n/2}\cos n\alpha}.$$

according as n is odd or even.

77. By equating the coefficients of n^2 in equation (4) of Art. 52, prove that

$$\frac{1}{6}\left(\sin^{-1}x\right)^2 = \frac{1}{2}\cdot\frac{x^2}{3} + \left(\frac{1}{1^2}+\frac{1}{3^2}\right)\frac{1\cdot3}{2\cdot4}\frac{x^5}{5} + \left(\frac{1}{1^2}+\frac{1}{3^2}+\frac{1}{5^2}\right)\frac{1\cdot3\cdot5}{2\cdot4\cdot6}\frac{x^7}{7}+...$$

78. Solve the equation sinh $x = 3x$ by a graphic solution, and thence obtain a nearer approximation by analytical methods.

79. Solve the equation $e^x = 3x$ by a graphic solution, and thence obtain nearer approximation to the answers by analytical methods.

80. If

$$\tan\frac{x}{2} = \tanh\frac{x}{2},$$

show that $\cos x \cosh x = 1$.

By means of a graph and the Tables, show that the smallest root of this equation is 4.730 approximately, and find roughly the values of the other roots.

81. If two of the roots of the cubic $x^3 - 3px + q = 0$ are imaginary, so that $q^2 > 4p^2$, show that (i) if p be negative, the roots are

$$\sqrt{-4p}\sinh u, \sqrt{-p}[-\sinh u \pm i\sqrt{3}\cosh u],$$

where

$$\sinh 3u = \frac{1}{2}\frac{q}{p\sqrt{-p}};$$

and (ii), if p be positive and q be negative, the roots are

$$\sqrt{4p}\cosh u, \sqrt{p}[-\cosh u \pm i\sqrt{3}\sinh u],$$

where

$$\cosh 3u = -\frac{1}{2}\frac{q}{p\sqrt{p}}.$$

82. If

$$x^{x^{x^{...\text{ad inf.}}}} = a(\cos\alpha + i\sin\alpha),$$

show that the general value of x is $r(\cos\theta + i\sin\theta)$,

where

$$\log r = \frac{(2n\pi + \alpha)\sin\alpha + \log a\cos\alpha}{\alpha},$$

and

$$\theta = \frac{(2n\pi + \alpha)\cos\alpha - \log a\sin\alpha}{\alpha}.$$

83. Criticise the fallacy

$$e^\theta = \left(e^{-\theta i}\right)^i = \left[e^{(2\pi-\theta)i}\right]^i = e^{\theta-2\pi}.$$

84. If

$$\frac{1}{\alpha} = (2n+1)\frac{\pi}{2},$$

where n has any positive integral value, prove that the solutions of the equation $\tan x = x$ are approximately

$$\pm\left(\frac{1}{\alpha} - \alpha - \frac{2\alpha^3}{3}\right).$$

85. Prove that $\dfrac{1}{1\cdot3} - \dfrac{1}{3\cdot5} + \dfrac{1}{5\cdot7} - \dfrac{1}{7\cdot9} + \ldots = \dfrac{\pi-2}{4}$,

and $\qquad\qquad \dfrac{1}{3\cdot5} - \dfrac{1}{7\cdot9} + \dfrac{1}{11\cdot13} - \dfrac{1}{15\cdot17} + \ldots = \dfrac{\pi\sqrt{2}-4}{8}$.

$$\left[\textit{For the second part, in the expansion of } \log\frac{1+x}{1-x} \textit{ put } x = i\sqrt{i} = \frac{i-1}{\sqrt{2}}.\right]$$

Sum to n terms the series:

86. $\dfrac{\sin 3\alpha}{\cos 2\alpha \cos 4\alpha} + \dfrac{\sin 5\alpha}{\cos 4\alpha \cos 6\alpha} + \dfrac{\sin 7\alpha}{\cos 6\alpha \cos 8\alpha} + \ldots$

87. $\dfrac{\tan^3\theta}{1-3\tan^2\theta} + \dfrac{1}{3}\dfrac{\tan^3 3\theta}{1-3\tan^2 3\theta} + \dfrac{1}{3^2}\dfrac{\tan^3 3^2\theta}{1-3\tan^2 3^2\theta} + \ldots$

88. $\dfrac{1+2\cos 2\alpha}{\sin 4\alpha} + \dfrac{1+2\cos 4\alpha}{\sin 8\alpha} + \dfrac{1+2\cos 8\alpha}{\sin 16\alpha} + \ldots$

89. $2\cos\dfrac{\theta}{2} + 2^2\cos\dfrac{\theta}{2}\cos\dfrac{\theta}{2^2} + 2^3\cos\dfrac{\theta}{2}\cos\dfrac{\theta}{2^2}\cos\dfrac{\theta}{2^3} + \ldots$

90. $\dfrac{2\cos\theta - \cos 3\theta}{\sin 3\theta} + 2\dfrac{2\cos 3\theta - \cos 3^2\theta}{\sin 3^2\theta} + \ldots + 2^{n-1}\dfrac{2\cos 3^{n-1}\theta - \cos 3^n\theta}{\sin 3^n\theta}$

91. $\dfrac{3\sin x - \sin 3x}{\cos 3x} + \dfrac{3\sin 3x - \sin 3^2 x}{3\cos 3^2 x} + \ldots + \dfrac{3\sin 3^{n-1}x - \sin 3^n x}{3^{n-1}\cos 3^n x}$

92. $\dfrac{\sin\alpha}{\cos 2\alpha} + \dfrac{\sin 3\alpha}{\cos 2\alpha \cos 4\alpha} + \dfrac{\sin 5\alpha}{\cos 4\alpha \cos 6\alpha} + \ldots$

93. $\dfrac{\sin\theta}{\cos\theta - \cos 2\theta} + \dfrac{\sin 3\theta}{\cos 3\theta - \cos 6\theta} + \dfrac{\sin 9\theta}{\cos 9\theta - \cos 18\theta} + \ldots$

94. $\dfrac{\sin 3x}{\sin x\sin^2 2x} + \dfrac{\sin 6x}{\sin 2x\sin^2 4x} + \dfrac{\sin 12x}{\sin 4x\sin^2 8x} + \ldots$

95. $\dfrac{1}{2}\tan 2\theta\tan^2\theta + \dfrac{1}{2^2}\tan 2^2\theta\tan^2 2\theta + \ldots + \dfrac{1}{2^r}\tan 2^r\theta\tan^2 2^{r-1}\theta + \ldots$

96. $\tan^{-1}\dfrac{12}{31} + \tan^{-1}\dfrac{12}{139} + \ldots + \tan^{-1}\dfrac{12}{36r^2-5} + \ldots$

97. $\tan^{-1}\dfrac{4}{7} + \tan^{-1}\dfrac{4}{19} + \ldots + \tan^{-1}\dfrac{4}{4r^2+3} + \ldots$

98. $\tanh^{-1}x + \tanh^{-1}\dfrac{x}{1-1\cdot2x^2} + \tanh^{-1}\dfrac{x}{1-2\cdot3x^2} + \ldots$

99. $\tan^{-1}\dfrac{x\sin\theta}{1+x\cos\theta} + \tan^{-1}\dfrac{x^2\sin 2\theta}{1+x^2\cos 2\theta} + \tan^{-1}\dfrac{x^4\sin 4\theta}{1+x^4\cos 4\theta} + \tan^{-1}\dfrac{x^8\sin 8\theta}{1+x^8\cos 8\theta} + ...$

100. $\cosh\dfrac{\theta}{2} + 2\cosh\dfrac{\theta}{2}\cosh\dfrac{\theta}{4} + ... + 2^{n-1}\cosh\dfrac{\theta}{2}\cosh\dfrac{\theta}{4}...\cosh\dfrac{\theta}{2^n}.$

101. $\dfrac{1}{2}\operatorname{cosec}\dfrac{\theta}{2}\cot\dfrac{\theta}{2} + \dfrac{1}{2^2}\operatorname{cosec}\dfrac{\theta}{2^2}\cot\dfrac{\theta}{2^2} + \dfrac{1}{2^3}\operatorname{cosec}\dfrac{\theta}{2^3}\cot\dfrac{\theta}{2^3} + ...$

102. Sum to infinity the series whose rth term is

$$\dfrac{1}{\underline{|r}}\cos r\theta\tan^r\theta.$$

103. Prove that

$$\theta - \sin\theta\cos\theta = 2\sin\theta\sin^2\dfrac{\theta}{2} + 2^2\sin\dfrac{\theta}{2}\sin^2\dfrac{\theta}{2^2} + 2^3\sin\dfrac{\theta}{2^2}\sin^2\dfrac{\theta}{2^3} + ... \text{ ad inf.}$$

104. Show that

$$\sum_{n=1}^{n=\infty}\dfrac{\tan\dfrac{\theta}{2^n}}{2^{n-1}\cos\dfrac{\theta}{2^{n-1}}} = \dfrac{2}{\sin 2\theta} - \dfrac{1}{\theta}.$$

105. Show that

$$\dfrac{\pi}{4} = \tan^{-1}\dfrac{1}{3} + \tan^{-1}\dfrac{2}{9} + \tan^{-1}\dfrac{4}{33} + \tan^{-1}\dfrac{8}{129} + ... \text{ to infinity,}$$

and that

$$\dfrac{\pi}{4} = \tan^{-1}\dfrac{1}{3} + \tan^{-1}\dfrac{2}{9} + \tan^{-1}\dfrac{4}{33} + \tan^{-1}\dfrac{8}{129} + \tan^{-1}\dfrac{1}{16}.$$

106. Show that

$$\sum_{n=1}^{n=\infty}\tan^{-1}\left(\dfrac{2}{n^2}\right) = \dfrac{3\pi}{4}.$$

107. Prove that

$$\cot^{-1}2 + \cot^{-1}8 + \cot^{-1}18 + \cot^{-1}32 + ... \text{ to } \infty = \dfrac{\pi}{4}.$$

108. Find the sum of the infinite series

$$1 + \dfrac{4}{\underline{|3}}x^3 + \dfrac{7}{\underline{|6}}x^6 + \dfrac{10}{\underline{|9}}x^9 + ...$$

109. Show that

$$x + \dfrac{x^7}{7} + \dfrac{x^{13}}{13} + \dfrac{x^{19}}{19} + ... \text{ ad inf.}$$

$$= \dfrac{1}{12}\log\dfrac{(1+x)^2(1+x+x^2)}{(1-x)^2(1-x+x^2)} + \dfrac{\sqrt{3}}{6}\tan^{-1}\dfrac{x\sqrt{3}}{1-x^2}.$$

110. Find the sum of series

$$1 - \frac{1}{7} + \frac{1}{9} - \frac{1}{15} + \frac{1}{17} - \frac{1}{23} + \frac{1}{25} - \dots \text{ ad inf.}$$

111. Prove that

$$\frac{1}{1^4 \cdot 3^4} + \frac{1}{3^4 \cdot 5^4} + \frac{1}{5^4 \cdot 7^4} + \dots \text{ to inf.} = \frac{\pi^4 + 30\pi^2 - 384}{768}.$$

112. Prove that

$$(1) \quad \frac{1}{1+4^2} + \frac{1}{1+8^2} + \frac{1}{1+12^2} + \frac{1}{1+16^2} + \dots \text{ ad inf.} = \frac{\pi}{8}\coth\frac{\pi}{4} - \frac{1}{2},$$

and that

$$(2) \quad \frac{1}{2^4 - 1} + \frac{1}{4^4 - 1} + \frac{1}{6^4 - 1} + \dots \text{ ad inf.} = \frac{1}{2} - \frac{\pi}{8}\coth\frac{\pi}{2}.$$

[*Use the result of Ex. 7, Page* 158.]

113. Prove that

$$\frac{1}{1^2 + x^2} - \frac{3}{3^2 + x^2} + \frac{5}{5^2 + x^2} - \dots \text{ ad inf.} = \frac{\pi}{4}\sec h\frac{\pi x}{2},$$

and that

$$\frac{1}{1^2 + x^2} + \frac{3}{3^2 + x^2} - \frac{5}{5^2 + x^2} - \frac{7}{7^2 + x^2} + \dots \text{ inf.}$$

$$= \frac{\pi}{2\sqrt{2}}\cosh\frac{\pi x}{4}\sec h\frac{\pi x}{2}.$$

[*Use the relation in Ex.* 9, *Page* 158, *substituting* $\frac{\pi}{4} - \frac{\theta}{2}$ *and* $\frac{\pi}{4} + \frac{\theta}{2}$ *for* θ.]

114. When n is even, prove that

$$\frac{(1+x)^n - (1-x)^n}{2x} = n \prod_{r=1}^{r=\frac{n}{2}-1} \left(x^3 + \tan^2\frac{r\pi}{n}\right).$$

Deduce that

$$\tan\frac{\pi}{2n}\tan\frac{2\pi}{2n}\tan\frac{3\pi}{2n}\dots \tan\frac{(n-1)\pi}{2n} = 1.$$

115. Prove that

$$\frac{(1+x)^n - (1-x)^n}{2x}$$

$$= \left(x^2 + \tan^2\frac{\pi}{n}\right)\left(x^2 + \tan^2\frac{2\pi}{n}\right)\dots\left(x^2 + \tan^2\frac{r\pi}{n}\right),$$

where $r = \frac{1}{2}(n-1)$, and n is odd.

116. Show that the infinite product

$$\frac{1+\dfrac{1}{1^2}}{1+\dfrac{1}{1\cdot 2}}\times\frac{1+\dfrac{1}{2^2}}{1+\dfrac{1}{2\cdot 3}}\times\frac{1+\dfrac{1}{3^2}}{1+\dfrac{1}{3\cdot 4}}\times\ldots$$

is equal to sech $\left(\dfrac{1}{2}\pi\sqrt{3}\right)\sinh\pi$.

117. If α, β, γ ... denote the prime numbers 2, 3, 5 ..., prove that

$$\frac{\alpha^2-\alpha^{-2}}{\alpha^2+1+\alpha^{-2}}\cdot\frac{\beta^2-\beta^{-2}}{\beta^2+1+\beta^{-2}}\ldots\text{ to infinity}=\frac{4}{7}.$$

118. Two regular polygons of n sides A_1, A_2, ... A_n, B_1, B_2, ... B_n are inscribed in the same circle of radius a. Prove that

$$\prod(A_r B_s)=2^n\,\alpha^{n^2}\,\sin^n\frac{n\theta}{2},$$

where r and s have all values from 1 to n, and θ is the angle between a pair of radii drawn one to a corner of each polygon.

119. $ABCD$ ... is a regular polygon of n sides, inscribed in a circle of radius a and centre O; show that the sum of the angles that AP, BP, CP ... make with OP is

$$\tan^{-1}\frac{a^n\sin n\theta}{a^n\cos n\theta-r^n},$$

where $OP=r$ and $\angle AOP=\theta$.

[*As in Art.* 119 *break* $x^n-a^n\cos n\theta+ia^n\sin n\theta$ *into linear factors, and apply the theorem of Ex.* 1 *Page* 193.]

120. Along a tangent to a circle are measured from the point of contact B distances BB_1, B_1B_2 ... equal to the diameter of the circle and C_1, C_2 ... are the middle points of these distances. From A, the other end of the diameter through B, lines are drawn to the points B_1, B_2 ... C_1, C_2 ... meeting the circle in b_1, b_2 ... c_1, c_2 ... Show that the product of the chords Bb_1, Bb_2 ... bears to the product of the chords Bc_1, Bc_2 ... the ratio

$$\sqrt{\pi\coth\pi}:1.$$

121. Show that

$$\tan^{-1}(\tanh y\cot x)=\tan^{-1}\frac{y}{x}-\sum_1^\infty\tan^{-1}\frac{2xy}{n^2\pi^2-x^2+y^2}.$$

Two points, P and Q, at a distance $2d$ apart are at the same distance c from a straight line and are also equidistant from one of an infinite series of points uniformly distributed along the line at distances a apart. Show the sum, θ, of the angles that PQ subtends at the points is such that

$$\tan\frac{\theta}{2}=\tan\frac{\pi d}{a}\coth\frac{\pi c}{a}.$$

[*Take* sin $(x+iy)$ *in factors* (*Art.* 122) *and apply the theorem of Ex.* 1, *Page* 193.]

122. Prove that

$$\prod_{n=1}^{n=\infty}\left[1-\frac{1}{n^2+n^{-2}}\right]=\frac{\cosh\pi-\cos\pi\sqrt{3}}{\cosh\pi\sqrt{2}-\cos\pi\sqrt{2}}.$$

[*In equation (2) of Art. 130 first put $2a = \pi$ and $2\theta = \pi\sqrt{3}$; next put $2a = \pi\sqrt{2}$ and $2\theta = \pi\sqrt{2}$. Divide one result by the other.*]

123. Show that the sum to infinity of the series

$$\tan^{-1} n^2 + \tan^{-1} \frac{n^2}{2^2} + \tan^{-1} \frac{n^2}{3^2} + \dots$$

is $\qquad \tan^{-1} \dfrac{\tan\alpha - \tanh\alpha}{\tan\alpha + \tanh\alpha}$, where $a = \dfrac{n\pi}{\sqrt{2}}$.

[*Start with the result of Art. 122 and put $\theta = n\pi\sqrt{-i}$.*]

124. Prove that

$$\tan^{-1} n^2 + \tan^{-1} \frac{n^2}{3^2} + \tan^{-1} \frac{n^2}{5^2} + \dots \text{ ad inf.} = \tan^{-1}(\tan\theta\tanh\theta),$$

where $\qquad \theta = \dfrac{n\pi}{2\sqrt{2}}$.

125. Prove that

$$\tan^{-1}(\cot\theta\coth\phi) + \tan^{-1}\left[\cot\left(\theta + \frac{\pi}{n}\right)\coth\phi\right]$$

$$+ \tan^{-1}\left[\cot\left(\theta + \frac{2\pi}{n}\right)\coth\phi\right] + \dots \text{ to } n \text{ terms}$$

is $\qquad \tan^{-1}(\cos n\theta\coth n\phi),$ or $\tan^{-1}(\cot n\theta\tanh n\phi),$

according as n is odd or even.

[*Use the result of Ex. 21, Page 145.*]

126. Find the sum of infinite series

$$\tan^{-1} \frac{1}{1^4} + \tan^{-1} \frac{1}{2^4} + \tan^{-1} \frac{1}{3^4} + \dots \text{ ad inf.}$$

[*Start with equation (2) of Art. 130 and put $2\alpha = 2\theta = \pi\sqrt{2} \cdot \sqrt{i}$.*]

127. Prove that

$$\tan^{-1} x - \tan^{-1} \frac{1}{3}x + \tan^{-1} \frac{1}{5}x - \dots \text{ ad inf.} = \tan^{-1}\left(\tanh\frac{\pi x}{4}\right).$$

$$\left[Put\ a = \frac{\pi}{4}\ and\ \theta = \frac{\pi x i}{4} in\ Ex.\ 13\ of\ Page\ 159.\right]$$

128. If x be a positive fraction, and if $\tan^{-1} y$ mean the least positive angle whose tangent is y, prove that

$$\sum_{r=0}^{r=\infty} (-1)^r \tan^{-1} \frac{(2r+1)x}{(2r+1)^2 - x^2} = \tan^{-1}\left[\sinh\frac{\pi x}{4}\sec\frac{\pi x\sqrt{3}}{4}\right].$$

129. If ABC be an acute-angled triangle, show that

$$\sin A + \sin B + \sin C > \cos A + \cos B + \cos C.$$

130. The internal bisectors of the angles A, B, C of a triangle meet the opposite sides at $D, E,$ and F. Show that the area of the triangle DEF cannot exceed one-fourth of the area of the triangle ABC.

131. If β lies between 0 and $\dfrac{\pi}{2}$, show that as θ increases from 0 to β the expression $\beta \sin \theta - \sin \beta$ first continually increases and then continually decreases.

132. Prove that the determinant

$$\begin{vmatrix} \cos\theta, & 1, & 0, & 0, & \cdots & 0 \\ 1, & 2\cos\theta, & 1 & 0, & \cdots & 0 \\ 0, & 1, & 2\cos\theta, & 1, & \cdots & 0 \\ \cdots & \cdots & \cdots & \cdots & \cdots & \cdots \\ 0, & 0, & \cdots & 1, & 2\cos\theta, & 1 \\ \cdots & \cdots & \cdots & 0, & 1, & 2\cos\theta \end{vmatrix}$$

of n rows and columns is equal to $\cos n\theta$.

133. Show that the nth convergent to the continued fraction

$$\frac{1}{2\tan\alpha} + \frac{1}{2\tan\alpha} + \frac{1}{2\tan\alpha} + \cdots$$

is

$$\frac{(\tan\alpha + \sec\alpha)^n - (\tan\alpha - \sec\alpha)^n}{(\tan\alpha + \sec\alpha)^{n+1} - (\tan\alpha - \sec\alpha)^{n+1}}.$$

134. Show that the nth convergent to the continued fraction

$$\frac{\sec^4\alpha}{2 - 2\tan^2\alpha} - \frac{\sec^4\alpha}{2 - 2\tan^2\alpha} - \frac{\sec^4\alpha}{2 - 2\tan^2\alpha} - \cdots$$

is

$$\frac{\sin 2n\alpha}{\cos^2\alpha \sin(2n+2)\alpha}.$$

135. Prove that

$$\frac{\sec^2\alpha}{4} - \frac{\sec^2\alpha}{1} - \frac{\sec^2\alpha}{4} - \frac{\sec^2\alpha}{1} - \cdots$$

to r quotients is equal to

$$\frac{\sin r\alpha}{2\sin(r+1)\alpha\cos\alpha}.$$

136. Show that

$$\frac{\tan\theta}{\theta} = \frac{1}{1} - \frac{\theta^2}{3} - \frac{\theta^2}{5} - \cdots$$

[In the following Examples it will be found convenient to use the Differential Calculus.]

137. Prove that

$$\cot^3\phi + \cot^3\left(\phi + \frac{\pi}{n}\right) + \cot^3\left(\phi + \frac{2\pi}{n}\right) + \cdots \text{ to } n \text{ terms}$$

$$= n^3 \operatorname{cosec}^2 n\phi \cot n\phi - n \cot n\phi.$$

[Differentiate twice the result of Ex. 6 of Page 73.]

138. The two equal sides of an isosceles triangle are given in length; show that when the radius of the inscribed circle is a maximum the angle between the equal sides is 76°, to the nearest degree.

139. Sum to n terms the series

$$\sec^2 x + \frac{1}{4}\sec^2 \frac{x}{2} + \frac{1}{4^2}\sec^2 \frac{x}{2^3} + \frac{1}{4^2}\sec^2 \frac{x}{2^3} + \ldots$$

140. Sum the series

$$\cos\theta + \frac{1}{2}\frac{\cos 3\theta}{3} + \frac{1\cdot 3}{2\cdot 4}\frac{\cos 5\theta}{5} + \ldots \text{ ad inf.}$$

141. Express $\dfrac{x^{n-1}}{x^{2n} - 2x^n \cos n\theta + 1}$ as the sum of n partial fractions with denominators quadratic in x.

142. Show that

$$\frac{\cos x}{\sin^3 x} = \frac{1}{x^3} + \frac{1}{(x+\pi)^3} + \frac{1}{(x-\pi)^3} + \frac{1}{(x+2\pi)^3} + \frac{1}{(x-2\pi)^3} + \ldots$$

[*Differentiate the result of Ex. 11, Page 159.*]

143. An infinite straight line is divided by an infinite number of points into portions each of length a. Prove that the sum of the fourth powers of the reciprocals of the distances of a point O on the line from all the points of division is

$$\frac{\pi^4}{3a^4}\left(3\operatorname{cosec}^4 \frac{\pi b}{a} - 2\operatorname{cosec}^3 \frac{\pi b}{a}\right),$$

where b is the distance of O from some one of the points of division.

[*Differentiate twice the result of Ex. 11, Page 159.*]

144. Prove that

$$\sum_{n=1}^{n=\infty} \frac{1}{n^4 + x^4} = \frac{\pi\sqrt{2}}{4x^3}\cdot\frac{\sinh \pi x\sqrt{2} + \sin \pi x\sqrt{2}}{\cosh \pi x\sqrt{2} - \cos \pi x\sqrt{2}} - \frac{1}{2x^4}.$$

[*In equation* (2) *of Art.* 130 *put* $2\alpha = 2\theta = \pi x\sqrt{2}$; *then take logarithms, and differentiate with respect to* x.]

Answers

Examples I. (Pages 7–9)

8. $\log_e 2$.

9. $\log_e 3 - \log_e 2$.

Examples II. (Pages 19–21)

1. $\sqrt{2}\left(\cos\dfrac{\pi}{4} + i\sin\dfrac{\pi}{4}\right)$.

2. $\sqrt{2}\left[\cos\left(-\dfrac{3\pi}{4}\right) + i\sin\left(-\dfrac{3\pi}{4}\right)\right]$.

3. $2\left[\cos\dfrac{5\pi}{6} + i\sin\dfrac{5\pi}{6}\right]$.

4. $5\left[\dfrac{3}{5} + i\cdot\dfrac{4}{5}\right]$.

5. $\sqrt{4+2\sqrt{2}}\left[\dfrac{\sqrt{2}+1}{\sqrt{4+2\sqrt{2}}} + i\dfrac{1}{\sqrt{4+2\sqrt{2}}}\right]$.

6. $(\sqrt{6}-\sqrt{2})\left[\cos\dfrac{5\pi}{12} + i\sin\dfrac{5\pi}{12}\right]$.

7. $\cos(10\theta + 12\alpha) - i\sin(10\theta + 12\alpha)$.

8. $\cos(\alpha + \beta - \gamma - \delta) + i\sin(\alpha + \beta - \gamma - \delta)$.

9. $\cos 107\theta - i\sin 107\theta$.

10. -1.

11. $\sin(4\alpha + 5\beta) - i\cos(4\alpha + 5\beta)$

12. $2^{n+1}\sin^n\dfrac{\theta-\phi}{2}\cos n\dfrac{\pi+\theta+\phi}{2}$.

23. $\cos\dfrac{\pi}{5} \pm i\sin\dfrac{\pi}{5}$; $\cos\dfrac{3\pi}{5} \pm i\sin\dfrac{3\pi}{5}$.

Examples III. (Pages 23–24)

1. 1; $\dfrac{-1\pm i\sqrt{3}}{2}$.

2. $\pm i$; $\dfrac{\sqrt{3}\pm i}{2}$; $\dfrac{-\sqrt{3}\pm i}{2}$.

3. $\pm\left(\cos\dfrac{r\pi}{12} + i\sin\dfrac{r\pi}{12}\right)$, where $r = 3, 7$, or 11.

4. $\pm i$, and $\pm\left(\cos\dfrac{r\pi}{10} \pm i\sin\dfrac{r\pi}{10}\right)$, where $r = 1$ or 3.

5. $\pm\sqrt[12]{2}\left(\cos\dfrac{r\pi}{24} + i\sin\dfrac{r\pi}{24}\right)$, where $r = 1, 9$, or 17.

6. $\sqrt[3]{2048}\left[\cos\dfrac{r\pi}{9} + i\sin\dfrac{r\pi}{9}\right]$, where $r = 5, 11$, or 17.

7. $\pm\sqrt[4]{2}\left[\cos\dfrac{r\pi}{12}-r\sin\dfrac{r\pi}{12}\right]$, where $r = 1$ or 7.

8. $\sqrt[3]{2}\left[\cos\dfrac{r\pi}{18}+i\sin\dfrac{r\pi}{18}\right]$, where $r = 1, 13$, or 25.

9. $\sqrt[5]{4}\left[\cos\dfrac{r\pi}{15}+i\sin\dfrac{r\pi}{15}\right]$, where $r = -1, 5, 11, 17$, or 23.

10. ± 2 and $\pm 2i$.

11. 2, and $2\left[\cos\dfrac{r\pi}{5}\pm i\sin\dfrac{r\pi}{5}\right]$, where $r = 2$ or 4.

12. -1024

13. $\pm\dfrac{i+\sqrt{3}}{2}$ and $\pm\dfrac{i\sqrt{3}-1}{2}$.

14. 1.

16. $\pm 1, \pm i, \pm\left(\cos\dfrac{\pi}{6}\pm i\sin\dfrac{\pi}{6}\right)$, and $\pm\left(\cos\dfrac{\pi}{3}\pm i\sin\dfrac{\pi}{3}\right)$.

The last four values.

17. -1 and $\cos\dfrac{r\pi}{7}\pm i\sin\dfrac{r\pi}{7}$, where $r = 1, 3$ or 5.

18. $-1, \cos\dfrac{\pi}{3}\pm i\sin\dfrac{\pi}{3}, \pm\left(\cos\dfrac{\pi}{4}+i\sin\dfrac{\pi}{4}\right)$, and $\pm\left(\cos\dfrac{3\pi}{4}+i\sin\dfrac{3\pi}{4}\right)$.

19. $2\sqrt[3]{2}\cos\dfrac{r\pi}{9}$, where $r = 1, 7$ or 13.

Examples IV. (Pages 28–29)

6. $\dfrac{5\tan\theta-10\tan^3\theta+\tan^5\theta}{1-10\tan^2\theta+5\tan^4\theta}.$

7. $\dfrac{7\tan\theta-35\tan^2\theta+21\tan^5\theta-\tan^7\theta}{1-21\tan^2\theta+35\tan^4\theta-7\tan^6\theta}.$

8. $\dfrac{9\tan\theta\quad 84\tan^3 0+126\tan^5\theta-36\tan^7\theta+\tan^9\theta}{1-36\tan^3\theta+126\tan^4\theta-84\tan^6\theta+9\tan^8\theta}$

Examples V. (Pages 36–38)

6. $3°48'51''.$

7. $\dfrac{1}{6}.$

8. $\dfrac{2}{m^2}.$

9. $\dfrac{a}{b}.$

10. $\dfrac{1}{2}.$

11. 3

12. $\dfrac{a^2}{b^2}.$

13. 0.

14. $-\dfrac{a^2+ab+b^2}{ab}.$

15. $-\dfrac{1}{2}$.

16. 2.

17. $-\dfrac{1}{6}$.

18. $-\dfrac{25}{14}$.

19. $-\infty$.

20. $2\dfrac{n^2 - m^2}{p^2}$.

21. $\dfrac{1}{60}$.

22. $\dfrac{2}{3}\dfrac{(m-n)^2}{mn}$.

23. 24.

24. 0.

25. $\log\dfrac{a}{b}$.

26. e.

27. e^2.

28. -9.

29. 1.

30. 0.

31. 1.

32. $e^{-\frac{x^2}{2}}$.

33. 0.

37. $\dfrac{8}{6}; -\dfrac{1}{6}$.

Examples VI. (Pages 41–42)

8. $x^5 - 55x^4 + 330x^3 - 462x^2 + 165x - 11 = 0$.

Examples IX. (Pages 58–59)

1. $\dfrac{1}{2^{n-1}}\cos n\theta$, $(n\ \text{odd})$; $\dfrac{1}{2^{n-1}}\left[(-1)^{n/2} - \cos n\theta\right]$, $(n\ \text{even})$.

2. $(-1)^{\frac{n-1}{2}}\dfrac{1}{2^{n-1}}\sin n\theta$, $(n\ \text{odd})$; $(-1)^{\frac{n}{2}}\dfrac{1}{2^{n-1}}(1 - \cos n\theta)$, $(n\ \text{even})$.

3. $n^2 \operatorname{cosec}^2 n\theta$, $(n\ \text{odd})$; $\dfrac{1}{2} n^2 \operatorname{cosec}^2 \dfrac{n\theta}{2}$, $(n\ \text{even})$.

4. $n^2 \sec^2 n\theta - n$, $(n\ \text{odd})$; $n^2 + \left[1 - (-1)^{\frac{n}{2}}\cos n\theta\right] - n$, $(n\ \text{even})$.

5. $-n\cot\left(\dfrac{n\pi}{2} + n\theta\right)$.

6. $n\cot n\theta$.

7. $(-1)^{\frac{n-1}{2}}\tan n\theta$, $(n\ \text{odd})$; $(-1)^{\frac{n}{2}}$, $(n\ \text{even})$.

8. $n^2 \cot^2\left(\dfrac{n\pi}{2} + n\theta\right) + n(n-1)$.

10. 0 or $\dfrac{1}{2}\dfrac{n^2}{(-1)^{\frac{n}{2}}\cos n\theta - 1}$, according as n is odd or even.

Examples XI. (Pages 68–70)

17. $\cos\alpha\cosh\beta - i\sin\alpha\sinh\beta$.

18. $\dfrac{\sin 2\alpha - i\sinh 2\beta}{\cosh 2\beta - \cos 2\alpha}$.

19. $2\dfrac{\sin\alpha\cosh\beta-i\cos\alpha\sinh\beta}{\cosh2\beta-\cos2\alpha}.$

20. $2\dfrac{\cos\alpha\cosh\beta+i\sin\alpha\sinh\beta}{\cos2\alpha+\cosh2\beta}.$

21. $\sinh\alpha\cos\beta+i\cosh\alpha\sin\beta.$

22. $\dfrac{\sinh2\alpha+i\sin2\beta}{\cosh2\alpha+\cos2\beta}.$

23. $2\dfrac{\cosh\alpha\cos\beta-i\sinh\alpha\sin\beta}{\cosh2\alpha+\cos2\beta}.$

Examples XII. (Pages 73)

1. $\pm\dfrac{\pi}{4}+\dfrac{i}{4}\log\dfrac{1+\sin\theta}{1-\sin\theta},$ according as $\cos\theta$ is positive or negative.

2. $\sin^{-1}(\sqrt{\sin\theta})+i\log[\sqrt{1+\sin\theta}-\sqrt{\sin\theta}].$

Examples XIII. (Page 78)

15. $\dfrac{1}{2}\log(u^2+v^2)+i\tan^{-1}\dfrac{v}{u},$ where

$u=\dfrac{1}{2}\log\dfrac{\cosh2y-\cos2x}{2},$ and $v=\tan^{-1}(\cot x\tanh y).$

Examples XV. (Pages 87–88)

1. 3   **2.** 2.

3. 5.   **4.** -1

5. -3.

Examples XVI. (Pages 91–92)

1. $\dfrac{4\sin\alpha}{5-4\cos\alpha}.$

2. 0, provided α does not equal a multiple of π.

3. $\dfrac{\sin^2\alpha}{1-\sin2\alpha+\sin^2\alpha}.$

4. $\dfrac{\sin\alpha(\cos\alpha-\sin\alpha)}{1-\sin2\alpha+\sin^2\alpha}.$

5. $\dfrac{\sin\alpha-c\sin(\alpha-\beta)-c^n\sin(\alpha+n\beta)+c^{n+1}\sin\{a+(n-1)\beta\}}{1-2c\cos\beta+c^2};\ \dfrac{\sin\alpha-c\sin(\alpha-\beta)}{1-2c\cos\beta+c^2}.$

6. $\dfrac{1-c\cosh\alpha-c^n\cosh n\alpha+c^{n+1}\cosh(n-1)\alpha}{1-2c\cosh\alpha+c^2}.$

7. $\dfrac{c\sinh\alpha}{1-2c\cosh\alpha+c^2}.$

8. $\dfrac{\cos\alpha+(-1)^{n-1}\{(n+1)\cos(n-1)\alpha+n\cos n\alpha\}}{2(1+\cos\alpha)}$.

9. $\dfrac{\sin\alpha+(2n+3)\sin n\alpha-(2n+1)\sin(n+1)\alpha}{2(1-\cos\alpha)}$.

10. 0, if $n = 4m$ or $4m + 3$, and 1, if $n = 4m + 1$ or $4m + 2$;
    0, if $n = 4m$ or $4m + 1$, and -1, if $n = 4m + 2$ or $4m + 3$;

11. $\left(2\cos\dfrac{\beta}{2}\right)^{n}\cdot\sin\left(\alpha+\dfrac{n\beta}{2}\right)$.

12. $(2\sin\alpha)^{-\frac{1}{2}}\sin\left(\dfrac{\pi}{4}+\dfrac{\alpha}{2}\right)$, except when $\alpha = n\pi$.

13. 0, if n be odd; $(-1)^{\frac{n}{2}}\sin^{n}\alpha$, if n be even.

14. $\left(2\sin\dfrac{\alpha}{2}\right)^{-n}\cdot\sin\left(\dfrac{n\pi}{2}-\dfrac{n\alpha}{2}\right)$, if n be < 1.

15. $\sqrt{\cos\theta(1+\cos\theta)}$, if θ be between $-\dfrac{\pi}{2}$ and $+\dfrac{\pi}{2}$.

16. $\left(2\cosh\dfrac{u}{2}\right)^{n}\cdot\sinh\dfrac{n+2}{2}u$.

Examples XVII. (Pages 94–96)

1. $e^{c\,\cos\beta}\sin(\alpha + c\sin\beta)$.   2. $e^{c\,\cos\beta}\cos(\alpha + c\sin\beta)$.

3. $e^{-\cos\alpha\,\cos\beta}\cos(\cos\alpha\sin\beta)$.

4. $\sin\alpha\cos(\cos\beta)\cosh(\sin\beta) - \cos\alpha\sin(\cos\beta)\sinh(\sin\beta)$.

5. $\sin(\cos\beta)\cosh(\sin\beta)\cos(\alpha - \beta) - \cos(\cos\beta)\sinh(\sin\beta)\sin(\alpha - \beta)$.

6. $e^{\cosh\alpha}\cosh(\sinh\alpha)$.

7. $e^{\cosh\alpha}\sinh(\sinh\alpha)$.

8. $e^{y\cos(\sin\alpha)}\cos\{y\sin(\sin\alpha)\}$, where $y = e^{\cos\alpha}$.

9. $e^{y\cos(\cos\alpha)}\cos\{y\sin(\cos\alpha)\}$, where $y = e^{\sin\alpha}$.

10. $\dfrac{1}{2}e^{\cos\theta}\{\cos(\theta + \sin\theta) + 4\cos(\sin\theta)\} + \dfrac{1}{2}e^{-\cos\theta}\{\cos(\theta - \sin\theta) - 4\cos(\sin\theta)\}$.

11. $\tan^{-1}\dfrac{c\sin\alpha}{1+c\cos\alpha}$, except when $c = 1$ and $\alpha = (2n + 1)\pi$.

12. $\dfrac{1}{2}\tan^{-1}\dfrac{2c\sin\alpha}{1-c^2}$, except when $c = 1$ and $\alpha = n\pi$.

13. $\dfrac{1}{4}\log\dfrac{1+2c\cos\alpha+c^2}{1-2c\cos\alpha+c^2}$.   14. $\dfrac{1}{2}\tan^{-1}\dfrac{2c\cos\alpha}{1-c^2}$.

15. $\dfrac{1}{4}\log\dfrac{1+2c\sin\alpha+c^2}{1-2c\sin\alpha+c^2}$.

16. $+\dfrac{\pi}{4}, -\dfrac{\pi}{4}$ or 0 according as cos α is positive, negative, or zero.

17. $\dfrac{1}{2}\cos(\alpha-\beta)\tan^{-1}\dfrac{2c\cos\beta}{1-c^2} - \dfrac{1}{2}\sin(\alpha-\beta)\tanh^{-1}\dfrac{2c\sin\beta}{1+c^2}.$

18. $\dfrac{1}{2}\log\left(\sin\dfrac{\alpha+\beta}{2}\mathrm{cosec}\dfrac{\alpha-\beta}{2}\right),$ except when $\alpha \pm \beta$ is a multiple of 2π.

19. $\dfrac{1}{2}\log\left[(1+c) \div \sqrt{1+2c\cos 2\alpha + c^2}\,\right].$

20. $\dfrac{\alpha}{2}.$
 21. $-\dfrac{1}{2}\tan^{-1}(\cos\beta\,\mathrm{cosech}\,\alpha).$

22. $\dfrac{1}{8}\left[2\sqrt{3}\log_c(2+\sqrt{3}) - \pi\right].$

Examples XVIII. (Pages 98–99)

1. $\cot\dfrac{\theta}{2} - \cot 2^{n-1}\theta.$
 2. $\mathrm{cosec}\,\theta\,\{\cot\theta - \cot(n+1)\theta\}.$

3. $\mathrm{cosec}\,\theta\,\{\tan(n+1)\theta - \tan\theta\}.$
 4. $\mathrm{cosec}\,\phi\,\{\tan(\theta+n\phi) - \tan\theta\}.$

5. $\dfrac{1}{2}\mathrm{cosec}\,\alpha\{\tan(n+1)\alpha - \tan\alpha\}.$

6. $S_n = \dfrac{1}{2^{n-1}}\cot\dfrac{\theta}{2^{n-1}} - 2\cot 2\theta; \; S_\infty = \dfrac{1}{\theta} - 2\cot 2\theta.$

7. $2\coth 2\theta - \dfrac{1}{2^{n-1}}\coth\dfrac{\theta}{2^{n-1}}.$
 8. $\tan 2^n\,\theta - \tan\theta.$

9. $\tan\theta - \tan\dfrac{\theta}{2^n}; \; \tan\theta.$
 10. $\sin\theta\,(\cot\theta - \cot 2^n\,\theta).$

11. $\dfrac{1}{2}\sin 2\theta + (-1)^{n+1}\dfrac{1}{2^{n+1}}\sin 2^{n+1}\theta.$
 12. $\dfrac{1}{2}\sin 2\theta - \dfrac{1}{2^{n+1}}\sin 2^{n+1}\theta.$

13. $\dfrac{1}{4}\mathrm{cosec}\dfrac{\theta}{2}\left(\sec\dfrac{2n+1}{2}\theta - \sec\dfrac{\theta}{2}\right).$
 14. $S_n = \dfrac{1}{2^{n-1}}\tan 2^n\,\alpha - 2\tan\alpha.$

15. $\dfrac{1}{4}\left\{3\cos\theta + \left(\dfrac{-1}{3}\right)^{n-1}\cos 3^n\theta\right\}.$
 16. $\dfrac{1}{4}\left\{3^n\sin\dfrac{\theta}{3^n} - \sin\theta\right\}.$

17. $\dfrac{1}{8}[3^n\tan 3^n\theta - \tan\theta].$
 18. $\dfrac{1}{2}[\cot\theta - 3^n\cot 3^n\theta].$

19. $\tan^{-1}\{(n+1)(n+2)\} - \tan^{-1}2.$

20. $\tan^{-1}(n+1) - \tan^{-1}1,$ *i.e.* $\tan^{-1}\dfrac{n}{n+2}.$

21. $S_n = \tan^{-1} 2^n - \tan^{-1} 1; \ S_\infty = \dfrac{\pi}{4}.$

22. $S_n = \sin^{-1} 1 - \sin^{-1} \dfrac{1}{\sqrt{n+1}}; \ S_\infty = \dfrac{\pi}{2}.$

Examples XIX. (Pages 102–103)

1. $1 - a \cos \theta + a^2 \cos 2\theta - a^3 \cos 3\theta + \ldots$ ad. inf.

2. $\cos \theta + a \cos (\theta + \phi) + a^2 \cos (\theta + 2\phi) + \ldots$ ad. inf.

3. $\sin \theta + a \sin (\theta + \phi) + a^2 \sin (\theta + 2\phi) + \ldots$ ad. inf.

4. $\cos\theta + a\cos(\theta + \phi) + \dfrac{a^2}{\underline{|2}}\cos(\theta + 2\phi) + \dfrac{a^2}{\underline{|3}}\cos(\theta + 3\phi) + \ldots$ ad inf.

5. $r\theta \sin \phi + \dfrac{r^2\theta^2}{\underline{|2}}\sin 2\phi + \dfrac{r^3\theta^3}{\underline{|3}}\sin 3\phi + \ldots$ ad inf.,

   where $r = +\sqrt{a^2 + b^2}$ and $\phi = \tan^{-1} \dfrac{b}{a}.$

9. $x\cos\alpha - \dfrac{1}{2}x^2 \sin 2\alpha - \dfrac{1}{3}x^3 \cos 3\alpha + \dfrac{1}{4}x^4 \sin 4\alpha + \dfrac{1}{5}x^5 \cos 5\alpha - \ldots$ ad inf.

10. $x + y - r\pi = -\cos\alpha \sin x - \dfrac{1}{2}\cos^2 \alpha \sin 2x - \dfrac{1}{3}\cos^2 \alpha \sin 3x - \ldots$ ad inf.

12. (1) $m = \tan^2 \dfrac{\alpha}{2};$ (2) $m = -\tan^2 \alpha.$

13. $-\log 2 - \sin 2\theta + \dfrac{1}{2}\cos 4\theta + \dfrac{1}{3}\sin 6\theta - \dfrac{1}{4}\cos 8\theta - \dfrac{1}{5}\sin 10\theta + \ldots$ ad inf.

14. $2\left[\sin\theta - \dfrac{1}{3}\sin 3\theta + \dfrac{1}{5}\sin 5\theta - \ldots \text{ ad inf.}\right]$

15. $2\left[\cos\theta \tan\left(\dfrac{\pi}{4} - \dfrac{\beta}{2}\right) - \dfrac{1}{2}\cos 2\theta \tan^2\left(\dfrac{\pi}{4} - \dfrac{\beta}{2}\right) + \ldots\right] + 2\log\cos\left(\dfrac{\pi}{4} - \dfrac{\beta}{2}\right),$ if $0 < \beta > \dfrac{\pi}{2}.$

Examples XX. [Pages 113–115]

1. $\Pi\left[x^2 + 2x\cos(3r + 1)\dfrac{2\pi}{9} + 1\right],$ where $r = 0, 1,$ or $2.$

2. $\Pi\left[x^2 - 2x\cos(6r + 1)\dfrac{\pi}{12} + 1\right],$ where $r = 0, 1, 2$ or $3.$

3. $\Pi\left[x^2 - 2x\cos(6r + 1)\dfrac{\pi}{15} + 1\right],$ where $r = 0, 1, 2, 3$ or $4.$

4. $\Pi\left[x^2 - 2x\cos(3r + 1)\dfrac{\pi}{9} + 1\right],$ where $r = 0, 1, 2, 3, 4$ or $5.$

5. $\Pi\left[x^2 - 2x\cos(6r+2)\dfrac{\pi}{21}+1\right]$, where $r = 0, 1, 2, 3, 4, 5$ or 6.

6. $(x-1)\Pi\left[x^2 - 2x\cos\dfrac{2r\pi}{5}+1\right]$, where $r = 1$ or 2.

7. $\Pi\left[x^3 - 2x\cos(2r+1)\dfrac{\pi}{6}+1\right]$, where $r = 0, 1,$ or 2.

8. $(x-1)\Pi\left[x^2 - 2x\cos\dfrac{2r\pi}{7}+1\right]$, where $r = 1, 2,$ or 3.

9. $(x+1)\Pi\left[x^2 - 2x\cos(2r+1)\dfrac{\pi}{9}+1\right]$, where $r = 0, 1, 2,$ or 3.

10. $(x^2-1)\Pi\left[x^2 - 2x\cos\dfrac{r\pi}{5}+1\right]$, where $r = 1, 2, 3,$ or 4.

11. $(x+1)\Pi\left[x^2 - 2x\cos(2r+1)\dfrac{\pi}{13}+1\right]$, where $r = 0, 1, ... 5$.

12. $(x^2-1)\Pi\left[x^2 - 2x\cos\dfrac{r\pi}{7}+1\right]$, where $r = 1, 2, ... 6$.

13. $\Pi\left[x^2 - 2x\cos(2r+1)\dfrac{\pi}{20}+1\right]$, where $r = 0, 1, 2, ... 9$.

29. Take the logarithm of both the sides of the expression of Art. 115 reading r instead of x; differentiate with respect to r and then integrate with respect to θ.

Examples XXII. (Pages 138–139)

2. $\pm 0.32746 ... m$, and $\pm 0.24989 ... m$.

3. $\dfrac{a\cos 2\beta}{\cos^2(\alpha+2\beta)}\delta$ and $\dfrac{a\sin^2\beta}{\cos^2(\alpha+2\beta)}\delta;$

$\dfrac{10\pi\sqrt{3}}{54}$ and $\dfrac{5(2-\sqrt{3})\pi}{54}$ metres.

7. $\dfrac{x - y\cos C}{c\sin B}$ and $\dfrac{y - x\cos C}{c\sin A}$ radians.

8. $-\dfrac{\pi}{480}$ metres.

Examples XXIII. (Page 143)

1. -1, and $\dfrac{1\pm\sqrt{3}}{2}$.

2. $-1 + 2\cos 40°$, $-1 + 2\cos 160°$, and $-1 + 2\cos 280°$.

3. -4, and $2\pm 2\sqrt{3}$.

4. 4, and $1\pm\sqrt{3}$.

5. $2\sqrt{7}\ \cos\theta$, where $\theta = 33°\ 37'\ 52''$, $153°\ 37'\ 52''$, and $273°\ 37'\ 52''$.

6. $-\dfrac{4}{3}+\dfrac{2\sqrt{10}}{3}\cos\theta$, where $\theta = 39°\ 5'\ 51''$, $159°\ 5'\ 51''$, and $279°\ 5'\ 51''$.

7. $\dfrac{2}{3}\sqrt{21}\cos\theta$, where $\theta = 44°\ 50'\ 49''$, $164°\ 50'\ 49''$, and $234°\ 50'\ 49''$.

Examples XXIV. (Pages 144–145)

3. The expression cannot lie between 2 and –2.

4. The least value is $\sqrt{a^2 - b^2}$, provided that $a > b$.

5. The greatest and least values are 3 and $\dfrac{1}{3}$ respectively.

6. The least value is $2ab$.

7. If a and b have the same signs, the least value is $(a + b)^2$.

8. $2 \tan \alpha$.

9. $2 \sec \alpha$.

Miscellaneous Examples XXV. (Pages 147–167)

4. $\dfrac{\pi}{2}; \pi.$

5. $2 - x \sin x - \cos x.$

6. $\dfrac{1}{x}+\dfrac{1}{3x^3}+\dfrac{1}{5x^5}+...$

9. $-2 - 4i; 1 \pm 2\sqrt{3} + i(2 \mp \sqrt{3}).$

10. $-2.$

12. $\dfrac{\pi}{4}\sin x$, if $0 < x < \dfrac{\pi}{2}$; 0, if $x = \dfrac{\pi}{2}$; $-\dfrac{\pi}{4}\sin x$, if $\dfrac{\pi}{2} < x < \pi$.

15. 1175 sq. metres.

20. $1 - \dfrac{\pi}{8}.$

21. $1 - (1 - \cos\theta)\log\left(2\sin\dfrac{\theta}{2}\right) - \dfrac{\pi - \theta}{2}\sin\theta$; unless θ be a multiple of 2π, when the sum is unity.

22. $\cot \alpha - 2^n \cot 2^n \alpha.$

23. $\dfrac{2}{\sin \pi}\left[\tan\dfrac{a}{2}\sin\theta + \tan^2\dfrac{a}{2}\sin 2\theta +...\right].$

24. $\dfrac{\pi}{(b-c)(c-a)(a-b)}[(Aa - B)(b - c)\cot \pi\alpha +...+...].$

31. $0.84...$

32. $\dfrac{3\sqrt{3}}{2}$

38. $\dfrac{1}{2}[\tan 3^n\theta - \tan\theta].$

40. $A.$

44. $\tan\theta - 2^n\tan\dfrac{\theta}{2^n}$.

45. $\dfrac{x}{\sin\theta}\cdot\sin\theta - \dfrac{1}{2}\dfrac{x^2}{\sin^2\theta}\sin 2\theta + \dfrac{1}{3}\dfrac{x^3}{\sin^3\theta}\sin 3\theta...$

47. $\tan\theta$

51. $\dfrac{\cos^2\alpha}{2\sin\alpha}\left[\dfrac{1}{\sin\alpha\sin 2\alpha} - \dfrac{1}{\sin(n+1)\alpha\sin(n+2)\alpha}\right]$.

52. $1 + \dfrac{1}{2}\cos\theta + \dfrac{1\cdot 3}{2\cdot 4}\cos 2\theta + \dfrac{1\cdot 3\cdot 5}{2\cdot 4\cdot 6}\cos 3\theta...$

54. $\dfrac{\pi^2}{3} - 3$.

60. $\cot\dfrac{\theta}{2^n} - \cot\theta$.

61. $\dfrac{\cosh(n+1)\alpha\cdot\sinh n\alpha}{\sinh\alpha}$.

78. 2.840.

79. 1.5121; 0.6191.

80. 7.8, 11.0; 14.1 ...

86. $\dfrac{1}{2}\operatorname{cosec}\alpha\,[\sec(2n+2)\alpha - \sec 2\alpha]$.

87. $\dfrac{1}{8}\left(\dfrac{1}{3^{n-1}}\tan 3^n\theta - 3\tan\theta\right)$.

88. $\sin 3\alpha\operatorname{cosec}\alpha\operatorname{cosec}2\alpha - \sin(3\,2^n\alpha)\operatorname{cosec}2^n\alpha.\operatorname{cosec}2^{n+1}\alpha$.

89. $\sin\theta\left[\cot\dfrac{\theta}{2^{n+1}} - \cot\dfrac{\theta}{2}\right]$.

90. $\cot\theta - 2^n\cot 3^n\theta$.

91. $\dfrac{1}{2}\left[\dfrac{1}{3^{n-1}}\tan 3^n x - 3\tan x\right]$.

92. $\sin^2 n\alpha\operatorname{cosec}\alpha\sec 2n\alpha$.

93. $\dfrac{1}{2}\left[\cot\dfrac{\theta}{2} - \cot\dfrac{3^n\theta}{2}\right]$.

94. $\operatorname{cosec}^2 x - \operatorname{cosec}^2 2^n x$.

95. $\dfrac{1}{2^n}\tan 2^n\theta - \tan\theta$.

96. $\tan^{-1}\dfrac{12n}{18n+13}$.

97. $\tan^{-1}\dfrac{4n}{2n-5}$.

98. $\tanh^{-1} nx$.

99. $\tan^{-1}\dfrac{x\sin\theta}{1-x\cos\theta} - \tan^{-1}\dfrac{x^{2n}\sin 2^n\theta}{1-x^{2^n}\cos 2^n\theta}$.

100. $\dfrac{1}{2}\sinh\theta\left[\coth\dfrac{\theta}{2^{n+1}} - \coth\dfrac{\theta}{2}\right]$.

101. $\dfrac{1}{2^{n+1}}\operatorname{cosec}^2\dfrac{\theta}{2^{n+1}} - \dfrac{1}{2}\operatorname{cosec}^2\dfrac{\theta}{2}$.

102. $e^{\sin\theta}\cos(\sin\theta\tan\theta) - 1$.

108. $\dfrac{1+x}{3}e^x + \dfrac{1}{3}e^{\frac{x}{2}}\left[(2-x)\cos\dfrac{\sqrt{3}x}{2} - \sqrt{3}x\sin\dfrac{\sqrt{3}x}{2}\right]$.

110. $\dfrac{\pi}{8}(\sqrt{2}+1)$.

126. $\tan^{-1}\dfrac{\sinh\lambda\sin\mu+\sin\lambda\sinh\mu}{\cosh\lambda\cos\mu-\cos\lambda\cosh\mu}-\dfrac{\pi}{4}$, where

$$\lambda=\pi\sqrt{2}\cos\frac{\pi}{8}\text{ and }\mu=\pi\sqrt{2}\sin\frac{\pi}{8}.$$

139. $4\csc^2 2x-\dfrac{1}{4^{n-1}}\csc^2\dfrac{x}{2^{n-1}}$.

140. $\dfrac{1}{2}\cos^{-1}(2\sin\theta-1)$; unless $\theta=n\pi$, when the sum is $\pm\dfrac{\pi}{2}$ according as n is even or odd.

141. $\dfrac{1}{n\sin n\theta}\displaystyle\sum_{r=0}^{r=n-1}\dfrac{\sin\left(\theta+\dfrac{2r\pi}{n}\right)}{x^2-2x\cos\left(\theta+\dfrac{2r\pi}{n}\right)+1}$.